# Studies of Brain Function, Vol. 13

# Studies of Brain Function

Yu. I. Arshavsky   I. M. Gelfand
G. N. Orlovsky

# Cerebellum and Rhythmical Movements

With 86 Figures

Springer-Verlag Berlin Heidelberg New York
London Paris Tokyo

Yu. I. Arshavsky
I. M. Gelfand
G. N. Orlovsky

Institute of Problems of
Information Transmission
Academy of Sciences of USSR
and Moscow State University
Moscow, USSR

Title of the original Russian edition:
Mozhechok i upravlenie ritmicheskimi dvizheniyami
© by Izdatelstvo NAUKA, Moscow, 1984

ISBN-13: 978-3-642-70830-5     e-ISBN-13: 978-3-642-70828-2
DOI: 10.1007/978-3-642-70828-2

Offsetprinting and binding: Konrad Triltsch, Graphischer Betrieb, Würzburg
2131/3130-543210

# Preface

After reading the manuscript, some biologists inquired why, on the basis of the broad experimental material presented in this book, we had not come up with a model describing the operation of the cerebellum. To answer this question, we decided to write a preface to our book.

How the nervous system copes with the complexity of the world is one of the central problems of neurophysiology. The question was clearly formulated for the first time by N.A. Bernstein. Considering the problem of motor control, he pointed out that the main objective of motor coordination is to overcome the redundant number of degrees of freedom of the motor apparatus or, in other words, to diminish the number of independent variables which control the movement (Bernstein 1967). These ideas were further developed by I.M. Gelfand and M.L. Zetlin (Gelfand and Zetlin 1966). They proposed, in particular, the "non-individualized" ("non-addressed") mode of control in complex systems, where only the highest levels of the system have the full notion about the final task while the main "effectors" act on the basis of very limited information. These propositions were made by Gelfand and Zetlin in a very general form, but, nevertheless, proved to be fruitful in determining the direction of experimental research. For instance, the discovery of the "locomotory region" of the brain stem (Shik et al. 1966a, b; 1967) was not a result of random experimenting, but was achieved by a deliberate search for a brain structure that, responding to a simple command, could evoke the complicated coordinated activity of a number of motor centres. The general ideas of Bernstein and "nonconcrete" models of Gelfand and Zetlin stimulated quite a number of interesting neurophysiological studies, for example, on postural regulation (see, e.g., Gurfinkel et al. 1965; Feldman 1979), on the control of rhythmical movements (see e.g., Orlovsky and Shik 1976; Shik and Orlovsky 1976; Berkinblit et al. 1978a, b) and others.

It is true that the "cybernetic vogue", which emerged in neurophysiology in the 1960's produced many concrete and, often, very ingenious models describing various functions of the brain (including the cerebellum). But just the concreteness of these models is their Achilles' heel. Firstly, this concreteness orients the researcher on solving narrow particular cases. Secondly, the experimentator organizes his work without references to the natural logic of the experiments, he begins to use the cybernetic language ("data flow", "bits of information", "controlled variables") and abandons the search of a language more adequate for the object studied. And thirdly, cybernetic models, especially those constructed for the nervous system of higher animals, are inevitably based on very limited experimental material. This gives a broad range for an arbitrary choice of the model's parameters and, consequently, the model turns out to be nonverifiable because one can explain any new experimental finding by changing some parameters of the model. This nonverifiability of a model by experiments brings cybernetic theories to stagnation, and a researcher can circle in vain for many years returning to the same ideas again and again. Easiness with which one can develop such theories (the material does not "fight back") casts a doubt on the fruitfulness of this way. We believe that the attempts to build a general theory of physiological processes based on cybernetic models have played their useful role but now this role is exhausted to a great extent. Authors of this book know this by first-hand experience. The construction of sufficiently verifiable conceptions is the main road of physiology. This road is as old as biology itself. No one can avoid this difficult, but fruitful way.

We do not advocate sheer empiricism, i.e., unlimited accumulation of facts without any general ideas. But to have a general idea in the form of a cybernetic model is indeed very dangerous. Since this book has a positive rather than discussive meaning, we intended to avoid the consideration of the existing theories of the cerebellum due to the reasons presented above. Such a discussion could lead us to the analysis of experimental evidenceses for or against every theory; this is out of the spirit of this book.

The experimental data described in this book were obtained in collaborative investigations carried out in the Institute of Problems of Information Transmission of the Academy of Sciences of the USSR and in the A.N. Belozersky Interfaculty Laboratory of Moscow State University. The authors are grateful to all of their colleagues who participated in the present study: L.I. Antziferova, M.B. Berkinblit, T.G. Deliagina, A.G. Feldman, O.I. Fukson, Yu.V. Panchin, G.A. Pavlova, C. Perret, L.B. Popova, M.L. Shik and V.S. Yacobson, as well as to E.G. Balzamo and Yu. Frolov for the assistance during the preparation of the English version of the book.

*Yu.I. Arshavsky*
*I.M. Gelfand*
*G.N. Orlovsky*

# Contents

# List of Abbreviations

| | | |
|---|---|---|
| EMG | – | Electromyogram |
| ENG | – | Electroneurogram |
| EPSP | – | Excitatory postsynaptic potential |
| IPSP | – | Inhibitory postsynaptic potential |
| S-phase | – | Short phase of the scratch cycle |
| L-phase | – | Long phase of the scratch cycle |
| DSCT | – | Dorsal spino-cerebellar tract |
| VSCT | – | Ventral spino-cerebellar tract |
| SRCP | – | Spino-reticulo-cerebellar pathway |
| SOCP | – | Spino-olivo-cerebellar pathway |
| bVFRT | – | Bilateral ventral flexor reflex tract |
| LRN | – | Lateral reticular nucleus |
| VS | – | Vestibulo-spinal |
| RS | – | Reticulo-spinal |
| RbS | – | Rubro-spinal |
| FN | – | Fastigial nucleus |
| IN | – | Interpositus nucleus |
| LN | – | Lateral nucleus |

# Introduction

A number of observations leave no doubt above the role of the cerebellum in motor control. (1) Removal of the cerebellum or its partial destruction results in motor disturbances. (2) Cerebellar output signals reach all motor centres of the nervous system: spinal cord, motor cortex, red and vestibular nuclei, reticular formation, etc. (3) The cerebellum receives signals from all motor centres as well as from the proprioceptors. (4) Stimulation of the cerebellum evokes various motor responses. (5) The activity of cerebellar neurons is correlated with movements.

However, the problem concerning the role played by the cerebellum in motor control remains unsolved. The fact is that cerebellar lesions do not result in elimination of any type of movements. "We know not a single function or reflex positively connected with the cerebellum in such a way that it is absent after cerebellar extirpation and present after ablation of other parts of the brain as long as the cerebellum remains uninjured" (Magnus 1925). Cerebellar disfunction results in deflections of different parameters which characterize movements (amplitude, force, speed, acceleration) from normal ones, in changes of relations between activities of various muscle groups, etc. (Luciani 1915; Holmes 1939; Dow and Moruzzi 1958). Thus, disturbances following cerebellotomy are usually called "discoordination", "decomposition", etc., and the cerebellum is regarded as an organ responsible for "coordinating", "regulating", and "smoothing" the movements. It was justly pointed out by Dow and Moruzzi (1958) that one should not overestimate the significance of such definitions. They do not clarify functions of the cerebellum, but rather emphasize the difference between the cerebellum and such specific motor centres as the spinal cord, oculomotor, respiratory, masticatory, deglutitory centres, the destruction of which leads to the loss of the corresponding movements.

Recently, a new approach in the study of the cerebellum was born: recording of cerebellar input and output signals immediately during movements. The structure of the cerebellum is highly favourable for such studies since both input and output neurons can be easily identified. This approach is based on two different methods. The first method was developed by Evarts (1966, 1968) and implies recording of the activity of single neurons in chronic experiments on intact animals trained to perform certain movements. This method was used while studying the activity of cerebellar neurons both in Evarts' laboratory and in other ones (Thach 1968, 1970a, b, 1972, 1975, 1978a, b; Grimm and Rushmer 1974; Mano 1974, 1979; Pauls et al. 1974; Mano and Yamamoto 1975, 1980; Robertson and Grimm 1975; Bioulac and Lamarre 1977; Burton and Onoda 1977, 1978; Gilbert and Thach 1977; Harvey et al. 1977, 1979; Soechting et al. 1978; J. Stein 1978; Strick 1983; Bauswein et al. 1980; Smith and Bourbonnais 1981; Yamamoto and Odagiri 1981). It is difficult to overestimate the value of the Evarts method which allows the recording of discharges of cerebellar neurons during complicated motor activity of an animal.

Another method, used in our studies and described in the present book, implies recording of the activity of single neurons in decerebrate animals performing "automatic" limb movements under conditions of acute experiments. The method appeared after development of "the preparation with controlled locomotion" (Shik et al. 1966a, b). This method, unlike the previous one, provides a good opportunity for analytical studies. On the other hand, in decerebrate animals, ony the brain stem-cerebellar and spinal motor mechanisms can be investigated; besides, one should take into account that removal of the cortex and other nervous centres might, to some extent, affect the activity of these mechanisms.

We studied the input and output signals of the cerebellum in decerebrate cats performing locomotor movements (more exactly, performing rhythmical stepping movements by the hindlimbs). We also studied these signals during the scratch reflex, in which one of the hindlimbs performs fast rhythmical oscillations. Simultaneously with this study, investigations of the spinal mechanisms of stepping and scratching movements were carried out in our laboratories. We hoped that progress in the study of the basic (spinal) mechanisms of rhythmical movements would promote understanding of the role of the cerebellum in the control of these movements.

Figure 1 shows schematically the main nervous mechanisms participating in the control of the hindlimb movements during locomotion and scratching in a decerebrate cat. Spinal mechanisms generating stepping and scratching movements are switched on by the signals arriving via special pathways either from the brain or from the upper segments of the spinal cord ("switching on"). When spinal mechanisms are operating and the cat performs stepping or scratching movements, various spino-cerebellar pathways convey signals on the activity of the spinal level of the system of motor control. These signals reach the

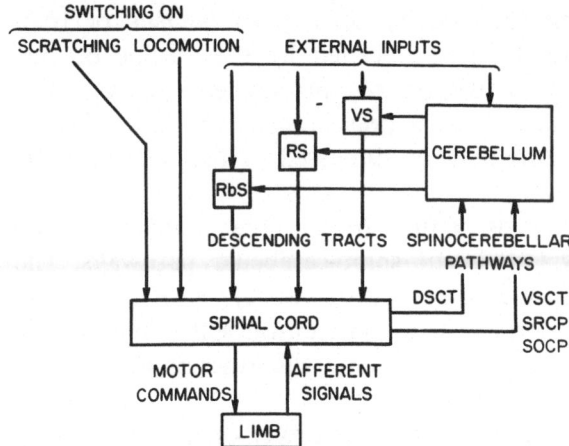

SWITCHING ON

SCRATCHING  LOCOMOTION        EXTERNAL INPUTS

**Fig. 1.** Spino-cerebellar loop. For explanations see text

medial and intermediate parts of the cerebellum and affect the cerebellar neurons which, in turn, influence the neurons of descending brain stem-spinal tracts. As a result, the descending tracts transmit the signals formed on the basis of information received by the cerebellum from the spinal cord. Thus, a closed chain ("spino-cerebellar loop") emerges, which comprises the spinal cord, spino-cerebellar pathways, cerebellum and descending tracts. Besides the signals from the "preceding" elements of the loop, both the neurons of descending tracts and the cerebellar neurons also receive information from other parts of the brain ("external inputs").

In the experiments which provided the material for this book, signals at different points of the spino-cerebellar loop during stepping and scratching movements were studied. For this purpose the activity of single identified neurons of the cerebellum, as well as of neurons giving rise to ascending and descending pathways was recorded. However, we are not trying to describe the activity of the system of motor control in terms of the activity of individual neurons. Even in invertebrates, whose nervous systems contain considerably less neurons, such a description seems to be hardly possible (except in the case of a few simple reflex acts) (cf. Kandel 1976; Selverston 1980). In the present book we use the data on the activity of individual cells only to obtain some integral characteristics of the activity of various parts of the system of motor control.

The book consists of seven chapters. Chapter I contains some information on the spinal mechanism controlling locomotion and scratching, as well as a description of the methods and preparations used.

The experimental data obtained by the authors of this book and their colleagues are described in Chaps. II-VI. The exposition of experimental facts concerning the activity of a given nucleus or tract during locomotion and scratching is preceded by the outline of anatomical and physiological organization of this nucleus (tract). Consideration of the experimental data is presented at the end of each chapter though some questions may be discussed in the course of exposition of experimental results.

Chapter II deals with signals coming to the cerebellum during locomotion and scratching via different spino-cerebellar pathways. The main experimental result is the discovery of two types of pathways: one conveys the signals on processes in the central spinal mechanism controlling stepping and scratching movements, the other on the activity of executive motor apparatus.

Chapter III describes the activity of neurons of the descending brain stem-spinal tracts, as well as the effects they produce on the spinal cord during rhythmical movements. The main conclusions made in this chapter are the following: the activity of descending tracts is modulated in relation to the rhythm of movements; this modulation is determined by the cerebellum; the signals arriving via the descending tracts affect the level of muscle activity, but not the phase of the activity in the cycle.

In Chapter IV the role of different spino-cerebellar pathways in generating the cerebellar output signals, i.e., in the rhythmical modulation of the neurons of descending tracts, is considered. It is demonstrated that the crucial role in generating cerebellar output signals is played by the information on the activity of the central spinal mechanism, whereas the information on the activity of the executive motor apparatus is of minor importance.

Chapter V deals with the activity of cerebellar neurons (Purkinje cells and neurons of the cerebellar nuclei). The role of different input signals in producing the rhythmical activity of cerebellar neurons is considered, as well as phase relations between the cerebellar and brain stem neurons.

In Chapter VI, the influence of one of the "external inputs", namely — the input from the vestibular apparatus — upon the activity of neurons of the vestibulo-spinal tract is considered. The main finding is that during scratching vestibular influence upon the spinal cord is modulated with the rhythm of scratching. In other words, the signals coming to the spinal cord depend both on the phase of the limb movement and on the position of the animal.

In Chapter VII, a hypothesis on the role of the medial and intermediate cerebellum in the control of stepping and scratching movements is advanced. We proceeded from the evident fact that in real life an animal often has to perform several motor acts at a time. For instance, during locomotion and scratching, it has to perform rhythmical limb movements and also to maintain the posture necessary for its equilibrium. While hunting, an animal has to combine locomotor movements with those movements which are involved in seizing the prey. Besides, the animal necessarily has to adapt its movements to the external conditions. For instance, while running in the forest, it has to correlate every step it makes with the signals coming from the visual system. We argue that the function of the cerebellum is that of coordinating different movements with each other, as well as correlating them with external conditions. The experimental material of Chaps. II-VI is used to support this hypothesis.

# I Spinal Mechanisms of Stepping and Scratching Movements

## 1. Crucial Role of the Spinal Cord in Control of Stepping and Scratching Movements

One of the first questions that arose in the process of studying nervous mechanisms of locomotion and scratching was the question on the role of the spinal cord and supraspinal structures in controlling these movements. At the end of the nineteenth century and at the beginning of the present one it was proven that under certain conditions stepping movements can be generated by the spinal cord (Freusberg 1874; Philippson 1905; Sherrington 1906a, 1910a; see reviews by Grillner 1973, 1975, 1981; Orlovsky and Shik 1976; Shik and Orlovsky 1976; Wetzel and Stuart 1976; P.G. Stein 1978). In chronic spinal dogs stepping movements of the hindlimbs may be evoked by exteroceptive or proprioceptive stimulation (Sherrington 1910a). But the muscular activity is weak, and the animal has to be kept above the ground. Stepping movements can be intensified by a small dose of strychnine (Hart 1971). Considerably better stepping can be observed in chronic cats in which the spinal cord is transected soon after birth (Shurrager 1955; Grillner 1973; Forssberg et al. 1975, 1980a, b). Their hindlimbs can bear the body weight. The animals can also vary the frequency of stepping in accordance with the speed of locomotion and they are capable of various gaits (alternating or in-phase stepping).

In acute experiments, stepping can be evoked by an injection of L-3,4-dihydroxyphenylalanine (DOPA) or Clonidine (Grillner 1969; Budakova 1973; Forssberg and Grillner 1973; Viala et al. 1974; Viala and Vidal 1978; Grillner and Zangger 1979; Zangger 1981). Exteroceptive stimulation considerably facilitates this process. In such experiments, the head of the animal is usually fixed while the body is either free or slightly suspended above the treadmill. The speed of the band determines the frequency of stepping and the type of gait; at high speed the limbs pass from alternating to in-phase stepping (gallop). In the best preparations, limb movements and the muscular activity pattern resemble those of intact animals.

Thus, the main features of stepping movements of the hindlimbs are determined by a mechanism located in the lumbo-sacral segments of the spinal cord. This mechanism can generate various rhythms corresponding to various speeds of locomotion; it can cause various degrees of muscle activities which result in more or less intense stepping; under certain conditions it can generate different gaits, i.e., different phase relations between the limbs. Finally, the spinal mechanism of stepping can, to some extent, adapt movements to external conditions. For instance, the limb can step over an obstacle, the weight of the animal being transferred to the symmetrical limb (Forssberg et al. 1975).

The fact that it is usually difficult to evoke intense movements in spinal preparations might simply mean that although the spinal mechanism is in principle capable of generating intense stepping, we have so far found no adequate means of its activation. Indeed, now when "loco-motor effects" of DOPA and Clonidine have been discovered, it may become possible to evoke more intense stepping in spinal preparations.

Let us now consider the role of the spinal cord in the scratch reflex. Sherrington demonstrated that the spinal cord can efficiently control scratching movements (Sherrington 1906b, 1910b). The hindlimb of a dog chronically spinalized in the middle of the thoracic region can easily "find" the stimulated point of the skin and scratch it. If the irritation is moved over the skin, the limb follows it. Evidently, the essential part of the nervous mechanism controlling scratching movements is located in the lumbo-sacral segments of the spinal cord.

These data suggest that in the lumbo-sacral segments of the spinal cord, there are neuronal mechanisms controlling stepping and scratching movements of the hindlimbs. The mechanisms can be switched on by the signals coming from higher motor centres. These signals (at least in locomotion) can be replaced by the action of certain drugs. Spinal mechanisms are quite efficient: (1) both stepping and scratching movements of the hindlimbs are well coordinated; (2) they are performed within a normal range of forces and joint angles; (3) in a number of cases, spinal mechanisms can adapt limb movements to varying external conditions.

## 2. Preparations and Evoking of Movements

To study the signals received and generated by the cerebellum during movements, in most experiments either mesencephalic or thalamic preparations with intact cerebellum were used. Only in a few cases, while studying the signals coming from the spinal cord, the experiments were carried out on decapitated preparations (transection of the spinal cord at the rostral border of the Cl segment).

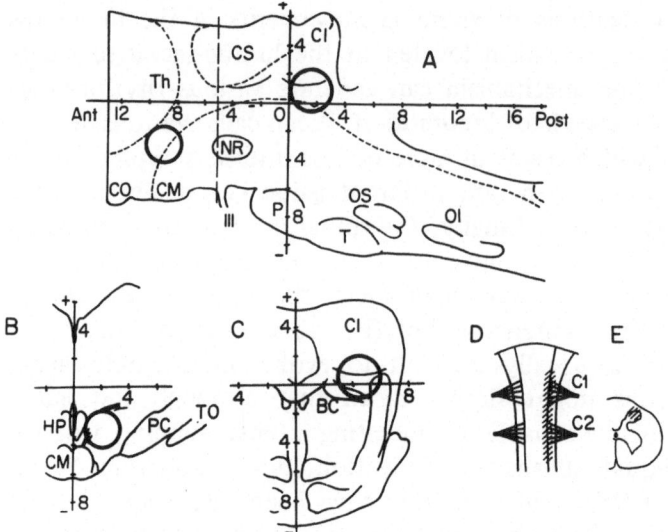

Fig. 2 A—E. Regions of the brain stem and of the spinal cord whose stimulation evokes rhythmical limb movements. **A** A parasagittal plane of the brain stem (about 2 mm of the midline), **B** A frontal plane A9 (Horsley Clarke's coordinates). **C** A frontal plane P2. The level of section (*A13*) to obtain the thalamic preparation and that (*A5, dashed line*) to obtain the mesencephalic preparation are shown in **A**. The *large circles* show the mesencephalic locomotor region; the *smaller circles* show the subthalamic one. **D** The rostral part of the spinal cord. **E** A cross-section of this part. *Hatched areas* in **D** and **E** show the region of the *C1* and *C2* segments whose stimulation evokes scratching movements. **Abbreviations:** *BC* brachium conjunctivum; *CI* colliculus inferior; *CM* mammillary bodies; *CO* optic chiasm; *CS* colliculus superior; *HP* hypothalamus posterior; *NR* nucleus ruber; *OI* inferior olive; *OS* superior olive; *P* pons; *PC* pedunculus cerebri; *T* trapezoid body; *Th* thalamus; *TO* tractus opticus; *III* third cranial nerve

Most studies of locomotion were carried out on the mesencephalic cat — "preparation with controlled locomotion" (Shik et al. 1966a, b). In order to obtain such a preparation, decerebration at the precollicular level was performed (marked by the vertical dashed line in Fig. 2A). The preparation is deprived of nociceptive perception; therefore, the skull and vertebral column can be rigidly fixed which is necessary for microelectrode studies. Though the mesencephalic preparation is incapable of spontaneous stepping movements, locomotion may be still easily evoked by stimulation of a certain area of the midbrain, which was called the mesencephalic locomotor region (large circle in Fig. 2A, C) (Shik et al. 1966a, b 1967).

The level of transection of the brain stem in the mesencephalic preparation coincides with the rostral border of the red nucleus (Fig. 2A). Thus, in the experiments in which the activity of neurons of the red nucleus was studied and in some other experiments, the thalamic preparation (transection of the brain stem at the level of the chiasma opticum, Fig. 2A) was used. Such a preparation is capable of spontaneous locomotion. Besides, locomotion may be evoked by stimulation of a certain area of the subthalamus (small circle in Fig. 2A, B) which was called the subthalamic locomotor region (Orlovsky 1969). A decorticated preparation used by Perret (1973) while studying the central and reflex mechanisms of locomotion resembles the thalamic one in many respects.

It remains unknown how stimulation of locomotor regions triggers the spinal mechanisms. Since, as it has been already mentioned, locomotion can be evoked by an injection of DOPA (the precursor of noradrenaline), it was suggested that the signals triggering the spinal mechanisms arrive via thin noradrenergic brain stem-spinal fibres (Grillner 1975). According to another hypothesis (Orlovsky 1970b), triggering of the spinal mechanisms generating stepping movements is produced by large reticulo-spinal neurons, since the beginning of locomotion is always preceded by an increase of the activity of these neurons (see Chap. III). Finally, a hypothesis was advanced that the triggering of the spinal locomotor mechanisms is performed through the chain of propriospinal neurons exciting each other in succession (Kazennikov et al. 1979, 1980).

Figure 3 shows the experimental arrangement used while studying locomotion. The cat's head, spine and pelvis are rigidly fixed, while the legs are left free. To make conditions for limb movements more natural, a treadmill with a moving band is placed under the cat so that the animal can walk or run on the band, depending on the strength of stimulation.

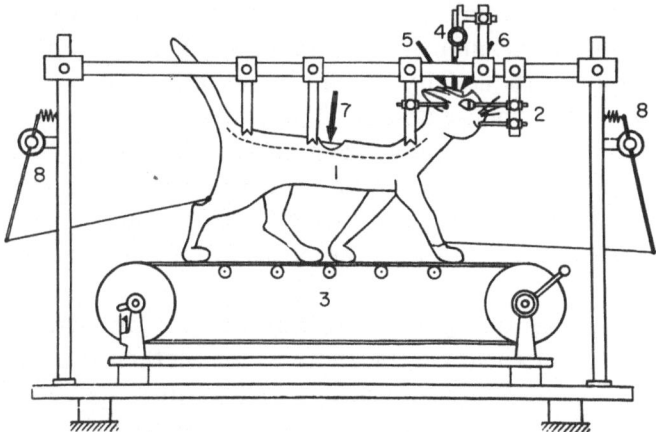

Fig. 3. Experimental arrangement. The thalamic or mesencephalic cat (1) is fixed in a sterotaxic device (2) with its limbs on the treadmill (3); electrodes (4–7) are inserted into the spinal cord and into the brain for stimulation and recordings. Limb movements are recorded by potentiometric transducers (8) (Shik et al. 1966b) 1966b)

In the experiments described in this book, the stimulation was of moderate strength, and the hindlimbs stepped in alternation as in a walk or trot. At such a strength, the forelimbs usually did not participate in locomotion. Stronger stimulation evoking gallop (when the hindlimbs move in-phase) was not performed.

To evoke scratching in decerebrate or thalamic cats, the method proposed by Sherrington (1910b, 1917) was used. In early experiments, scratching was evoked by electrical stimulation of upper cervical segments (C1-C2) of the spinal cord (hatched area in Fig. 2D, E). In most experiments scratching was evoked by tactile stimulation of the pinna (Deliagina et al. 1975). To facilitate the scratch reflex, d-tubocurarine (0.2-0.5%) was applied to the dorsal surface of the C1-C2 segments (Domer and Feldberg 1960; Feldberg and Fleischhauer 1960).

Signals from upper cervical segments triggering the spinal mechanism of scratching, which is located in the lumbo-sacral segments, seem to be conveyed through a special propriospinal pathway (Berkinblit et al. 1977; Deliagina 1977). This pathway crosses the midline in the upper cervical region and then, in the thoracic region, returns to the original side.

The experimental arrangement for studying the scratch reflex is practically the same as that for locomotion (Fig. 3); naturally, there is no need for a treadmill.

A considerable part of our experiments was devoted to studies of the scratch reflex in animals immobilized with Flaxedil (Deliagina et al. 1975). Later it will be demonstrated that the activity of the spinal mechanism of scratching scarcely depends on the afferent signals coming from receptors of the moving limb. Thus, this mechanism can also be easily triggered in immobilized animals by stimulation of the receptive field of the scratch reflex. Its activity results in rhythmical bursts of motoneurons typical of normal scratching. The activity of the spinal mechanism of scratching in an immobilized animal was called "the fictitious scratch reflex" since it is not accompanied by any limb movements. Fictitious scratching can be evoked only if the hindlimb is passively deflected far forward, i.e., put into the position typical of actual scratching.

Immobility of an animal during fictitious scratching[1] considerably facilitates the work with microelectrodes. Besides, the method of fictitious scratching enables one to study the role of afferent signals in cerebellar activity. It has been mentioned above that the activity of the spinal

---

[1] We shall use the terms "fictitious scratching" and "scratching" indifferently to denote fictitious scratch reflex

mechanism of scratching hardly changes after an animal has been immobilized. On the other hand, during fictitious scratching there are no
rhythmical signals from limb receptors. Thus, comparing the activity
of different neurons, nuclei or tracts during actual and fictitious scratching one can estimate the role of the central and peripheral (afferent)
signals in their activity.

## 3. Hindlimb Movements and Muscle Activity

The movements of the hindlimb and activity of its muscles during treadmill locomotion of a decerebrate cat (Gambarian et al. 1971; Orlovsky
and Shik 1976) are very close to natural ones (Engberg and Lundberg
1969; Rasmussen et al. 1978), and only minor differences can be found
(Wetzel et al. 1975). It is convenient to divide the whole step cycle
into two parts – the stance phase, when the limb touches the ground,
and the swing phase (Fig. 4). In the stance phase, a group of extensor
muscles is active. While extending the limb joints, these muscles counteract
body weight and develop a propulsive force for moving the animal
forwards (or, under the experimental conditions shown in Fig. 3, accelerate the treadmill band). Not only "pure" extensors, but also some
two-joint muscles (which flex one joint and extend another) are active
in the stance phase. Movements in joints are also schematically shown in
Fig. 4. The hip joint is extending throughout the stance phase because
of contraction of its extensors, and the hip (as well as the whole limb)
is moving backwards relative to the body. Knee and ankle movements

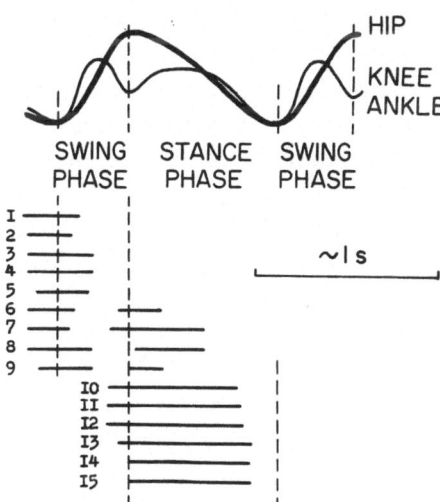

Fig. 4. The joint movements and muscular activity during stepping movements
of the hindlimb. *Curves* deflect upwards
when joints are flexing; muscles
illustrated are: *1* rectus femoris;
*2* tensor fascia latae; *3* tibialis anterior;
*4* extensor digitorum longus; *5* iliopsoas;
*6* biceps pars posterior; *7* gracilis;
*8* sartorius; *9* semitendineus; *10* soleus;
*11* gastrocnemius, plantaris; *12* vastus;
*13* adductor femoris, semimembranosus;
*14* biceps pars anterior; *15* gluteus.
Time calibration is presented for the
average step cycle
(After Gambarian et al. 1971)

are more complicated. During the first half of the stance phase these joints are slightly flexing, in spite of the counteraction of the active extensors, which is considered to be a "yield" to the body weight (Lundberg 1969; Grillner 1972). Then, by the very end of the stance phase, all joints are extending because of active extensor contraction and of movement of the animal relative to the supporting surface.

When the limb reaches the most caudal position, the swing phase (i.e., protraction of the limb forwards) begins. Both flexors and some two-joint muscles become active at the end of the stance phase, which results in flexion of all limb joints after cessation of the stance phase. The flexion terminates earlier in the knee and ankle joints, and they begin to extend. This extension seems to be determined by a decrease of flexor activity, by onset of weak extensor activity and by inertial forces which act on the distal parts of the limb (Engberg and Lundberg 1969; Lundberg 1969; Gambarian et al. 1971). All limb extensors have been strongly activated shortly (about 20 ms) before the limb touches the ground (Engberg and Lundberg 1969). Therefore, the limb is capable of supporting the body weight from the very beginning of the stance phase.

This sequence of activities of various muscles as well as the sequence of movements in various joints is, for the most part, preserved both during weak and intense locomotion. Some changes are observed in two-joint muscles, which can exhibit one or two bursts of activity during the step cycle, depending on the intensity of locomotion (Gambarian et al. 1971; Perret and Cabelguen 1976, 1980).

Recordings of the activity of motoneurons during locomotion showed that they discharge with frequencies of 30-50 pulses $s^{-1}$ during the periods of activity of the corresponding muscles, and that they are inhibited between the bursts of activity (Shik et al. 1966a; Severin et al. 1967b; Perret and Cabelguen 1980; Zajac and Young 1980; Perret 1983).

The study of activity of muscles during locomotion demonstrated that the pattern of this activity in the step cycle is very complex; it can be considered as reciprocal, alternating activity of the flexor and extensor groups only as a rough approximation (Sherrington 1910a; Lundberg 1969). Indeed, only because of differences in activities of hip flexors and of flexors of other joints, can the complex limb movement in the swing phase be realized (Perret and Cabelguen 1980).

Nevertheless, we shall use a traditionally simplified pattern of muscle activity (i.e., reciprocal, alternating activities of flexors and extensors) while considering neuronal mechanisms controlling movements. The beginning of the swing phase will be usualy taken as the beginning of the step cycle.

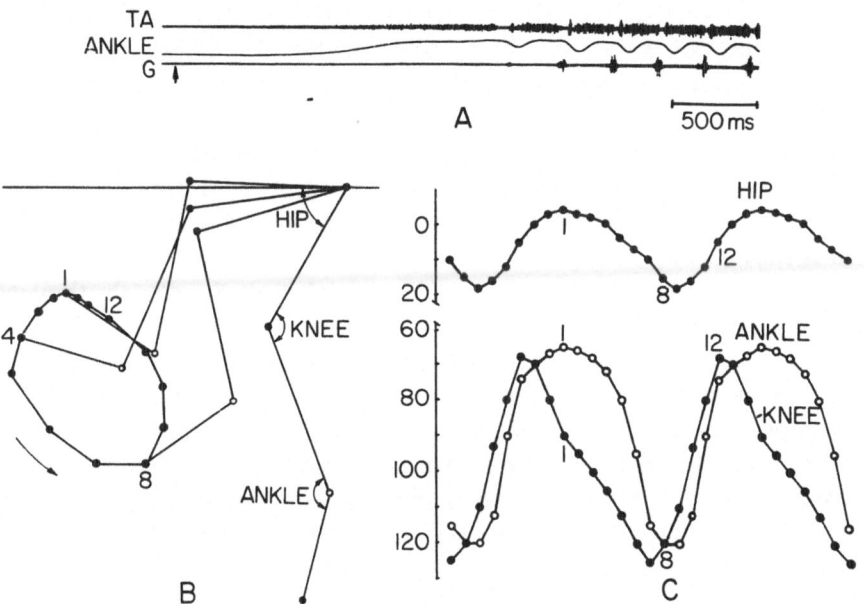

**Fig. 5 A−C.** Movements of the hindlimb and activity of its muscles during scratching. **A** Angular movements in the ankle joint recorded together with EMG's (electromyogram) of m.tibialis anterior (*TA*) and m.gastrocnemius (*G*). **B** An initial position of the limb and three positions during scratching (*1, 4* and *8*). Successive positions of the distal point of the foot are connected by *lines* to present a trajectory of this point. Angular movements in various joints are shown in **C**, *numbers* indicate the limb position given in **B**. Intervals between *dots* are 20 ms, the angular scale (degrees) is on the *left* in **C**. (The data presented in **B** and **C** were obtained by filming) (Deliagina et al. 1975)

Let us now consider movements of the hindlimb and activity of its muscles during scratching. With stimulation of the pinna (or of the upper cervical cord) the limb joints are strongly flexing (especially the hip and ankle) (Fig. 5A), and the limb protracts far forward. This is the postural stage of the scratch reflex. Joint flexions are determined by the tonic activity of flexors; activity of one of them (m.tibialis anterior) is shown in Fig. 5A. Then a rhythmical stage of the reflex begins: the limb rhythmically oscillates with a period of 250 ms (the right part of Fig. 5A). The oscillations are determined by short bursts of the activity of extensors and some two-joint muscles; activity of one of the extensors (m. gastrocnemius) is shown in Fig. 5A. Each burst of the extensor activity is accompanied by the relaxation of flexors.

In natural conditions, during the postural stage of the scratch reflex, the limb approaches the stimulated area of the skin and then starts scratching it (Sherrington 1906b). In our experiments the cat, being fixed, could not bend the body towards the limb; thus, scratching movements were performed "in the air", i.e., the limb did not touch the skin.

Figure 5B shows the trajectory of the distal point of the foot as well as three instantaneous positions of the limb (1, 4, 8) during scratching; the initial limb posture is also shown. One can see how far forward the limb is moved, i.e., how great the postural component of the movement is. Figure 5C shows movements in different joints during the rhythmical stage of scratching.

The pattern of muscle activity, which corresponds to this motor pattern, is rather complicated (Deliagina et al. 1975, O'Donovan et al. 1982). Since most experiments in neuronal mechanisms of scratching were carried out on immobilized preparations, we shall describe the pattern of the efferent activity in muscle nerves during fictitious scratching (Deliagina et al. 1981). One can see from Fig. 6 that the nerves which become

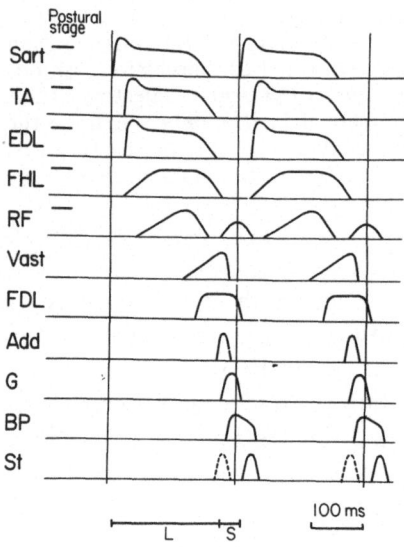

**Fig. 6.** The efferent activity in various muscle nerves during fictitious scratching. *Vertical lines* show the onset of the activity in n.sartorius. At the foot of the graph, a scratch cycle consisting of the long (*L*) and short (*S*) phases is shown; the termination of the n.gastrocnemius activity is assumed to be the beginning of the cycle, the onset of its activity — the beginning of the S-phase. On the *left*, the nerves with a tonic activation during the postural stage of the reflex (i.e., during the latent period of rhythmical scratching) are marked. The *dashed line* shows the second, not regularly observed, burst in the *St* nerve. Time calibration is presented for the average scratch cycle. **Abbreviations:** *Sart* sartorius; *TA* tibialis anterior; *EDL* extensor digitorum longus; *FHL* flexor hallucius longus; *RF* rectus femoris; *Vast* vastus; *FDL* flexor digitorum longus; *Add* adductor femoris; *G* gastrocnemius; *BP* biceps pars posterior; *St* semitendinosus (After Deliagina et al. 1981)

active in the "postural" stage of the reflex (in nonimmobilized animals they determine the postural component of movement) are active not exactly in-phase. Similarly, short bursts in the extensor nerves do not appear simultaneously, but are shifted in relation to each other and to pauses in the activity of postural nerves. Thus, the pattern of muscle activity during scratching can be considered as reciprocal and alternating activity of the flexor and extensor groups (Sherrington 1910b) only as a rough approximation, which is also the case for locomotion. Nevertheless, we shall make use of this simplified picture of efferent activity while considering nervous mechanisms controlling scratching. The scratch cycle is assumed to comprise two phases — the short one (S) when n.gastrocnemius is active, and the long one (L) when it is not. The termination of n.gastrocnemius burst is taken as the cycle beginning, and the onset of this burst as the transition between the L- and S-phases. As it follows from Fig. 6, if this definition of the phases of the scratch cycle is accepted, activity of most flexors (sartorius, tibialis anterior, extensor digitorum longus, rectus femoris) falls mainly into the L-phase, while activity of most extensors (gastrocnemius, vastus, adductor femoris) — into the S-phase; some muscles are active between the L- and S-phases as biceps posterior (Deliagina et al. 1981).

Figure 7 presents intracellular records from the flexor (A-C) and extensor (D) motoneurons during fictitious scratching (Berkinblit et al. 1980). During the postural stage, the flexor motoneuron becomes depolarized (B). During rhythmical generation, the motoneuron is depolarized and fires in the L-phase of the scratch cycle being periodically inhibited in the S-phase (C). In the extensor motoneuron, depolarization gradually increases during most of the L-phase and reaches the maximum in the S-phase when the motoneuron fires (D). The firing rate in flexor motoneurons is the same as during locomotion (30-50 pulses $s^{-1}$), while in extensor motoneurons it is considerably higher (100-300 pulses $s^{-1}$).

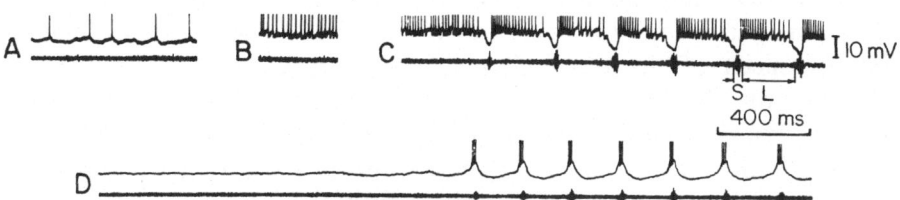

Fig. 7 A—D. Intracellular recording of the activity of flexor and extensor motoneurons during fictitious scratching. A—C Activity of the motoneuron of m.tibialis anterior at rest (A), during the latent period of rhythmical scratching (B) and during scratching (C). D Activity of the motoneuron of m.gastrocnemius. The *lower trace* is the gastrocnemius ENG (electroneurogram). L- and S-phases of the cycle are indicated in C (Berkinblit et al. 1980)

## 4. Role of Central Mechanisms and Afferent Signals in Generating Rhythmical Movements

On of the principal questions arising from the study of locomotion and scratch reflex is that of the role of central mechanisms and afferent signals sensory feedback) in generating the rhythmical activity underlying these movements. As shown by Sherrington (1910b), a deafferented limb is capable of scratching movements. Comparison of joint movements and muscle activity of intact and deafferented limb has shown that the basic pattern of scratching movements (their rhythm and phases of activity of flexors and extensors) usually persists after deafferentation. Cessation of the afferent inflow to the spinal cord from limb receptors results only in a change of the level of muscle activity (Deliagina et al. 1975). Nor does the efferent pattern of fictitious scratching in an immobilized animal differ from that of actual scratching (Deliagina et al. 1975, 1981). Thus, the rhythm of scratching and phases of activity of various muscles in the cycle are determined by the central spinal mechanism. One may observe that the rhythm of scratching is strikingly constant. By varying the strength of stimulation of the pinna or of the cervical spinal cord one can considerably change the intensity of scratching, but hardly affect its rhythm.

The efferent pattern of stepping movements is, to some extent, also determined by the central spinal mechanisms. Under certain conditions stepping movements can be evoked in a completely deafferented preparation (Grillner and Zangger 1979). In an immobilized preparation, "fictitious locomotion" with rhythmical efferent activity typical of actual locomotion can also be evoked (Viala and Buser 1969; Perret 1973, 1976, 1983; Zangger 1978; Grillner and Zangger 1979; Jordan et al. 1979; Vidal et al. 1979; Perret and Cabelguen 1980). However, in contrast to scratching, rhythm and other parameters of activity of the central spinal mechanisms controlling stepping movements are strongly dependent upon sensory feedbacks. In controlled locomotion, duration of a locomotor cycle can vary within a wide range depending on the speed of the treadmill band (Shik et al. 1966a). Correlation between the rhythm of stepping and the speed of the treadmill band may be explained as follows. In the stance phase of the step the limb is moving backwards in relation to the body. It is the time necessary for the limb to move from the rostral position to the caudal on that determines duration of the stance phase which is the largest part of the step duration. The length of steps being approximately the same, the time necessary for reaching the caudal position depends on the speed of the band and can vary within a wide range (Shik et al. 1966a). When the limb reaches the caudal position, some proprioceptors inform the spinal cord. As a result, the flexors are activated and the swing phase begins (Pearson and Duysens 1976; Andersson et al. 1978; Grillner and Rossignol 1978).

Thus, during both locomotion and scratching a central mechanism generating rhythmical process is triggered in the spinal cord. However, in scratching, the activity of this mechanism depends only to a small extent upon sensory feedback; this mechanism determines the rhythm of movements as well as flexor and extensor phases of the cycle. During locomotion, activity of the central mechanism to a considerably greater extent depends upon afferent signals which affect the rhythm. Due to this dependence, a new cycle of activity of the spinal generator can begin only after the previous mechanical cycle is fulfilled. Up to this moment, afferent signals coming from the moving limb prevent activation of the mechanism responsible for protraction of the limb forward (swing phase). If the limb is deafferented, this "veto" does not exist and the spinal generator can function as a truly central one, i.e., it can determine its own rhythm of oscillations. In such a case, the rhythm generated is usually within the range of rhythms typical of normal stepping (Orlovsky and Feldman 1972a; Feldman and Orlovsky 1975; Grillner and Zangger 1979).

It should be pointed out, that the normal activity of the spinal mechanism of stepping and its adaptation to the varying speed of the treadmill band requires only a very weak rhythmical afferent inflow. When deafferenting the hindlimbs, it is sufficient to leave only one-half of the L7 dorsal root intact to provide this inflow (Orlowsky and Feldman 1972a). After complete deafferentation of the hindlimbs, the rhythm of their movements is determined by afferent signals from the forelimbs (Shik et al. 1966b). In immobilized animals, the rhythm of fictitious locomotion can be strongly affected even by weak passive limb oscillations (few degrees): the rhythm of the oscillations fully determines the rhythm of the spinal generator (Andersson et al. 1978).

The locomotor rhythm generated by the spinal mechanism depends not only on afferent signals, but also on supraspinal influences. For instance, during fictitious locomotion, the rhythm can be considerably changed by varying the strength of stimulation of the locomotor region (Feldman and Orlovsky 1975; Grillner and Zangger 1979).

## 5. Afferent Signals

A great number of receptors of various kinds provide the spinal cord with information on limb movements. Among them, only the signals from the primary endings of muscle spindles (conveyed by Ia fibres) were thoroughly studied during locomotion (Severin et al. 1967a; Severin 1970; Perret and Buser 1972; Perret, Berthoz 1973; Sjöström and Zangger 1976; Feldman et al. 1977; Prochazka et al. 1976, 1977; Loeb and Duysens 1979; Perret and Cabelguen 1980; Cabelguen 1981). Unlike other proprioceptors, muscle spindles are directly influenced by the central nervous system: they are innervated by axons of gamma-motoneurons (see Granit 1970; Matthews 1972). Thus, the activity of spindle afferents is determined by two factors, i.e., by the muscle length and by the activity of the corresponding gamme-motoneurons. It was found that in various movements activity of gamma-motoneurons is "linked" to that of alpha-motoneurons resulting in excitation of the spindle afferents in the phase of active muscle contraction. Figure 8 shows the activity of a primary spindle afferent (group Ia) from the ankle extensor muscles during locomotion (Severin et al. 1967a). Stretching of these muscles with passive ankle flexion results in activation of the afferent (A). During locomotion of low intensity (B) the afferent is active both in the stance phase of the step, when ankle extensors are actively contracting, and in the swing phase, when they are passively stretched. With more intense locomotion (C) activity of the afferent in the stance phase increases in-parallel with extensor activity, while its discharge in the swing phase disappears in spite of increased muscle stretching in

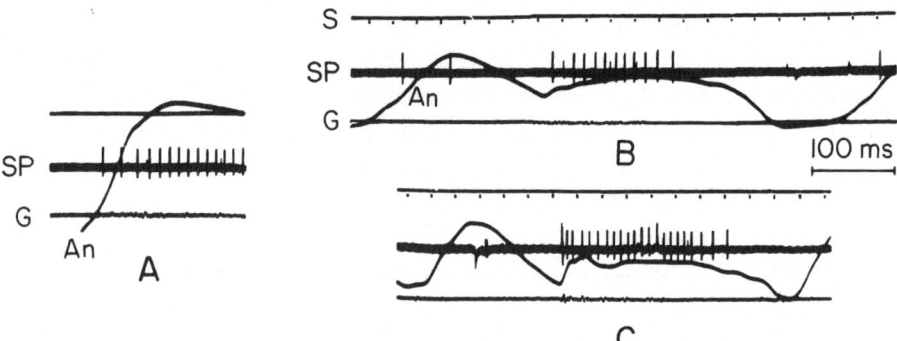

**Fig. 8 A—C.** Activity of a muscle spindle afferent from the ankle extensor. **A** Response to the passive ankle flexion. **B, C** Activity during weak (**B**) and intense (**C**) locomotion. Activity of the afferent (*SP*) is recorded together with the ankle angle (*An*) and gastrocnemius EMG (*G*). *S* stimulation of the locomotor region (Severin et al. 1967a)

this phase. This might be accounted for by the fact that gamma-moto-neurons are coactivated with alpha-motoneurons; as a result, spindle afferents become active when the corresponding muscles are actively contracting. Direct recordings from gamma-motoneurons have shown that they are really generating bursts of impulses with the rhythm of stepping (Severin 1970). Similar behaviour in spindle afferent was observed during scratching (Feldman et al. 1977).

Besides the signals from muscle spindle afferents, the signals from Golgi tendon organs have also been recorded during controlled locomotion (Severin et al. 1967a). As it could be expected, these receptors were active in the phase of contraction of the corresponding muscle when it develops the maximum force.

## 6. Localization of the Rhythmical Generator. Activity of Neurons of the "Leading" Region of the Lumbo-Sacral Spinal Cord

Neuronal mechanisms controlling movements of the hindlimbs during stepping and scratching are located in the spinal hindlimb centres, i.e., in the lumbo-sacral enlargement (the L3–S1 segments). Some data exist suggesting that this region is functionally nonhomogeneous. In chronic experiments on cats it was demonstrated that the capability of generating stepping movements persisted in caudal spinal segments after transection of the spinal cord above the L4 segment while after transection below L4 stepping could not be evoked (Afelt et al. 1973). In acute experiments with administration of DOPA as a "stimulator" of locomotion, rhythmical generation could be evoked even after transection of the spinal cord in the rostral part of L5; after transection in the caudal part of L5, generation could not be evoked (Grillner and Zangger 1979). These findings suggest that the role of rostral segments of the lumbosacral enlargement in generating the rhythm of stepping is vitally important.

Corresponding results were also obtained while studying the role of these segments in scratching. Stable generation of the scratching rhythm could be evoked in the rostral lumbar segments after transection of the spinal cord between L5 and L6 (Berkinblit et al. 1978a) and after functional "switching-off" of the L5 and more caudal segments by means of cooling (Deliagina et al. 1983). A scheme of the experiments with the cooling of the spinal cord is shown in Fig. 9A, B. When the cold water flows through a thermode applied to the cord (A, B), neurons of the cooled spinal segment located on the side of the thermode lose the ability of spike discharge (C–E), and the conductance through

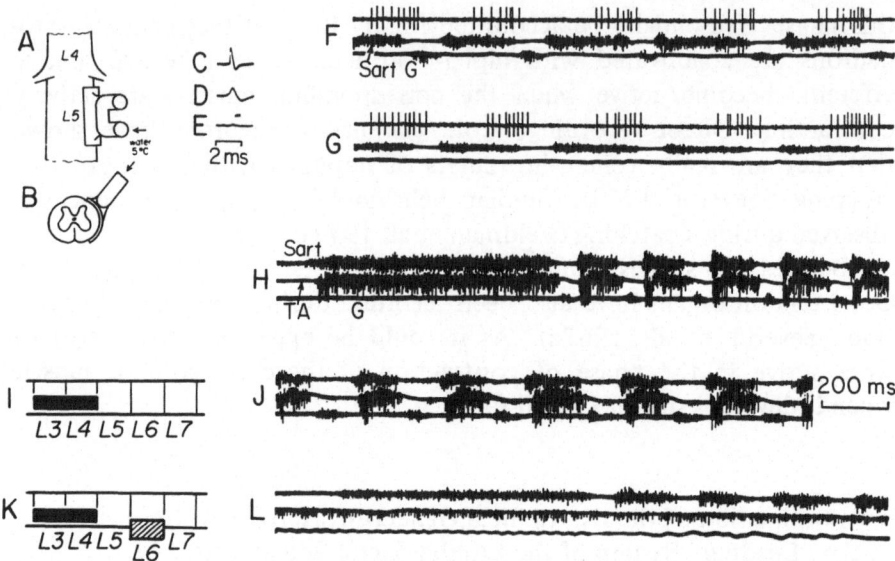

**Fig. 9 A—L.** Effects of cooling of the spinal cord upon generation of the scratch rhythm. **A, B** Experimental arrangement — the metal thermode is applied to the lateral surface of the L5 segment. **C—E** The effects of cooling upon the action potentials of an interneuron from *L5*. Recordings were performed before cooling **(C)**, 30 s **(D)** and 60 s **(E)** after the beginning of cooling. **F, G** Activity. of an interneuron from *L4* recorded together with ENG's of n.sartorius (*middle trace*) and n.gastrocnemius (*lower trace*) during fictitious scratching before cooling of the *L5* segment **(F)** and 60 s after the beginning of cooling **(G)**. **H—J** Effects of destruction of the spinal cord grey matter in the *L3* and *L4* segments. **I** A scheme of the destruction. The *black rectangle* shows the area of unilateral destruction. Fictitious scratching before destruction is shown in **H**, that after destruction in **J. K, L** Generation of the scratch rhythm by an isolated *L5* segment. A scheme of the experiment is shown in **K**. The grey matter of the *L3* and *L4* segments was destroyed, and the caudal part of the cord was inactivated by the thermode applied to the *L6* segment (*hatched rectangle*). Under these conditions, pinna stimulation evoked a normal temporal pattern of scratching in n.sartorius **(L)**. In **H, J** and **L** the ENG's of n.n.sartorius (*Sart*), tibialis anterior (*TA*) and gastrocnemius (*G*) are presented (Deliagina et al. 1983)

axons in the white matter is blocked. Thus, the cooling of a given spinal segment results in "switching-off" of neuronal mechanisms of the cooled segment and of more caudal ones. Figure 9F, G shows the effect of cooling of the L5 segment. In this experiment, the activity of neuronal mechanisms in the rostral region of the lumbo-sacral enlargement was determined from the activity of an interneuron located in L4 as well as from the activity of n.sartorius (the sartorius motoneuron pool is located in L4 and L5, Romanes 1964). The activity of neuronal mechanisms of caudal segments was determined from the activity of n.gastrocnemius that originates in L7 and S1. During cooling, the caudal part of the spinal cord is functionally switched off: the rhythmical activity in n.gastrocnemius disappears and that in n.sartorius decreases (Fig. 9G). Nevertheless, cooling of the L5 segment affects neither the rhythm of scratching nor the pattern of rhythmical activity of the interneuron from L4; both the bursts position in the cycle and the firing rate within the burst persist during cooling.

Thus, inactivating the L5 and more caudal segments, i.e., the greater part of the spinal hindlimb centre, practically does not affect the process of rhythmic generation in the rostral part of the lumbo-sacral enlargement (in the L3 and L4 segments).

Not only the L3 and L4 segments are capable of rhythmical generation. This was proven in the experiments in which the grey matter of the L3 and L4 segments was destroyed by means of thermocoagulation while the main pathways descending to the caudal part of the spinal cord were preserved (Deliagina et al. 1983). Figure 9H shows activity of nerves to m.m.sartorius, tibials anterior and gastrocnemius during the fictitious scratch reflex. Then the grey matter of the L3 and L4 segments was destroyed (Fig. 9I) which produced no marked effect upon the rhythmic activity (Fig. 9J). Thus, not only the rostral part of the lumbo-sacral enlargement (the L3 and L4 segments), but also its caudal part (the L5 and more caudal segments) are capable of rhythmical generation.

The described experiments have demonstrated that the capacity for generation of rhythmical oscillations, presenting the basis for rhythmical scratching movements, is distributed along the lumbo-sacral spinal cord. However, this distribution is not even: the capacity is most pronounced in the rostral part of the lumbo-sacral enlargement, i.e., in the L3—L5 segments. The capacity of this region is so pronounced that even its parts — the L3 and L4 segments (Fig. 9G) as well as the isolated L5 segment (Fig. 9K, L) — can generate the stable scratching rhythm. More caudal segments (L6—S1) are less capable of rhythmic generation. It seems likely that the L3—L5 segments are "leading" ones, i.e., that they determine the rhythm of activity in the whole spinal hindlimb centre. On the other hand, the caudal part of the lumbo-sacral enlargement (the L6—S1 segments), in which the capability for rhythmical generation is weak or is absent, contains the main "output" spinal mechanisms, i.e., most hindlimb motoneurons and most interneurons of various reflex chains. These data together with those of Afelt et al. (1973) and of Grillner and Zangger (1979), considered at the beginning of this section, suggest that the rostral part of the lumbosacral enlargement specializes in the generation of rhythmical activity underlying different rhythmical movements of the hindlimbs.

To obtain data about the organization of the rhythmical generator, activity of interneurons from the L4 and L5 segments was studied

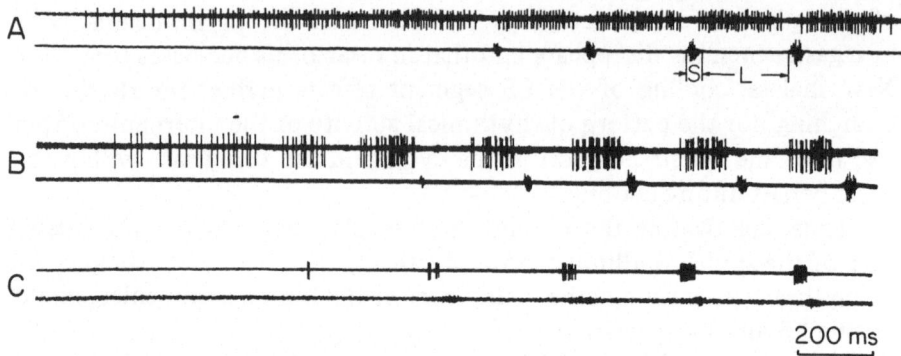

Fig. 10 A–C. Activity of spinal interneurons of the group 1 (A), group 2 (B) and group 3 (C) during fictitious scratching. The *lower trace* is the ENG of n.gastrocnemius. The *L*- and *S*-phases of the cycle are indicated in A (Berkinblit et al. 1978a)

during fictitious scratching (Berkinblit et al. 1978a). Interneurons were recorded throughout the spinal grey cross-section except for the region of the motor nuclei. During fictitious scratching, the discharge of a great many interneurons is modulated with the rhythm of motoneuronal activity. Three examples of rhythmically active neurons are presented in Fig. 10. The neurons fire in bursts and are silent between bursts. Various neurons are active in various phases of the scratching cycle: Fig. 10A shows a neuron firing throughout the greater part of the L-phase; Fig. 10B shows a neuron firing in the second half of the L-phase; Fig. 10C shows the neuron firing at the end of the L-phase and at the beginning of the S-phase. Rhyhmically active neurons are located in Rexed's (1954) layer VII, and to a smaller extent in V, VI and VIII layers.

Figure 11 shows the timing of bursts of interneurons in the normalized scratch cycle ("phase distribution"). The cycle duration is assumed to be a unit, the transition between the L- and S-phases is marked by an interrupted line. Curves F and E show approximate dependence between the membrane potential of flexor and extensor motoneurons and the phase of the cycle (Berkinblit et al. 1980). The burst position of each neuron in the normalized cycle is presented by a horizontal line. In this graph, the neurons are organized according to the onset of the burst: the later the neuron begins to fire in the cycle, the lower it is presented on the graph. This order is broken only for a few neurons that begin firing at the very end of the cycle, which are shown in the upper part of the graph.

As one can see in Fig. 11, phases of neuron "switching on" are distributed rather evenly all over the cycle, i.e., new neurons are continously recruited throughout the cycle. Burst terminations ("switching off" of the neurons) are distributed mainly in the second half of the cycle.

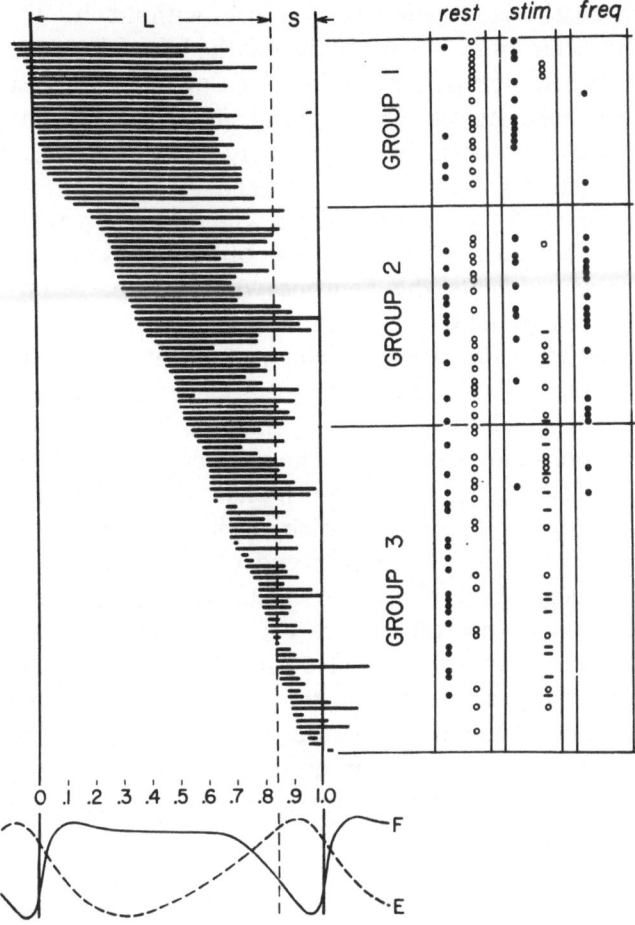

**Fig. 11.** Phase distribution of 120 spinal interneurons during fictitious scratching and table of their paramters. *Solid vertical lines* show beginning and end of the normalized cycle, a *dashed line* — transition between *L*- and *S*-phases. The burst position of each neuron in the normalized cycle is shown by a *horizontal line*. Curves *F* and *E* show an approximate dependence between the membrane potential of flexor and extensor motoneurons and the phase of the cycle. In the *rest* column, the behaviour of 83 neurons under resting conditions is shown (*filled circles*, neurons with resting discharge; *open circles*, without resting discharge). In the *stim* column, the behaviour of 52 neurons during the latent period of scratching is shown (*filled circles*, neurons are activated; *open circles*, neurons have no resting discharge and are not activated; *minuses*, resting discharge of neurons is inhibited). In the *freq* column, *filled circles* indicate neurons whose discharge frequencies increase in the course of the burst (Berkinblit et al. 1978a)

Thus, the number of simultaneously active neurons changes considerably during the cycle: it increases during the first half of the cycle, reaches a maximum near the middle of it and then decreases (Fig. 12A).

Figure 11, besides the phase distribution, presents a table of some other parameters of the neurons. In the *rest* column, filled circles indicate neurons with resting discharge, open circles those without it. The *stim* column shows the behaviour of neurons during the latent period of rhythmical generation (when stimulation of the pinna or of the cervical spinal cord has already been started, but rhythmical oscillations have not yet appeared). Neurons which were tonically activated during this period (as neurons in Fig. 10A, B) are marked by filled circles; neurons which were inhibited by minuses; those which had no resting discharge and were not activated (Fig. 10C) by open circles. In the *freq* column, filled circles indicate the neurons whose discharge rate increased in the course of the burst (as in a neuron in Fig. 10B).

In order to find correlations between the phase of activity and other parameters of interneurons, all the units were arbitrarily divided into three groups (Fig. 11). Group 1 comprises the units which begin firing at the very end of the preceding cycle or at the very beginning of the given one and become silent at the end of the L-phase (as a neuron in Fig. 10A). Group 2 comprises the units which begin firing near the middle of the L-phase and stop firing at the end of this phase (as a neuron in Fig. 10B). Finally, group 3 comprises the units active only at the end of the cycle (as a neuron in Fig. 10C). The three groups are marked in Fig. 11. One can see distinct correlations between the phase of activity and other parameters of the neurons. Firstly, the relative amount of spontaneously active neurons is highest in group 3, somewhat lower in group 2 and lowest in group 1. Secondly, during the latent period of rhythmical generation, the group 1 neurons usually become active while the resting discharge of the group 3 neurons is

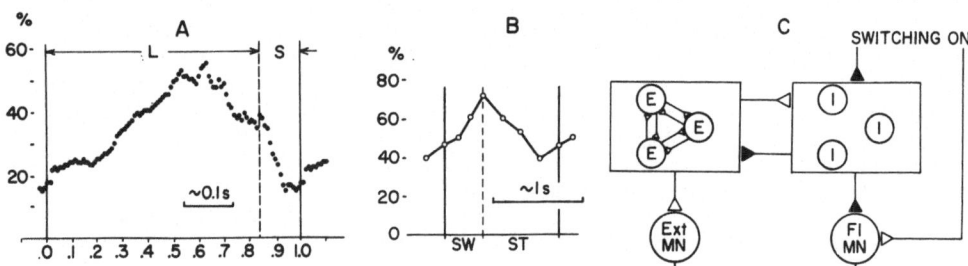

**Fig. 12A, B.** Relative number of active spinal interneurons in different phases of the cycle. **A** Fictitious scratch reflex (Berkinblit et al. 1978a). **B** Locomotion of the cat with deafferented hindlimbs; *SW* and *ST* are swing and stance phases of the locomotor cycle (Orlovsky and Feldman 1972b). **C** Model of the generator of the scratching rhythm; excitatory and inhibitory synapses are shown by *white* and *black* triangles, respectively; *Ext MN* and *Fl MN* are extensor and flexor motoneurons

inhibited. Thirdly, an increase of the discharge frequency in the course of the burst is typical of the group 2 neurons. In the next section we shall consider the possible functional role of different groups of interneurons.

Activity of interneurons was also studied during locomotion of mesencephalic cats with deafferented hindlimbs (Orlovsky and Feldman 1972b). In this case, the behaviour of interneurons is similar to that during fictitious scratching. The amount of active units increases during the swing (flexor) phase and decreases during the stance (extensor) phase of the locomotor cycle (Fig. 12B).

## 7. Conclusion

The data described in Chap. I prove that rhythmical processes underlying stepping and scratching movements are generated by the central spinal mechanisms. In scratching, the activity of the central mechanism scarcely depends on sensory feedbacks. On the contrary, activity of the central locomotor mechanism is considerably affected by afferent signals coming from receptors of the moving limbs. The study of other "automatic" movements — such as mastication, swallowing, eye nistagmus — has revealed that their basic patterns can also be generated by central mechanisms in absence of sensory feedback (Sumi 1969, 1970; Dellow and Lund 1971; Miller 1972; Lund 1976; Hikosaka et al. 1977). For a long time it has also been known that the respiratory rhythm can be generated in the medulla oblongata in the absence of the afferent inflow from the respiratory system (Gesell et al. 1936; Baumgarten 1956). Finally, various movements in invertebrates are also controlled by central generators since corresponding efferent patterns can be obtained in an isolated nervous system or in a single ganglion (see Kandel 1976; Kennedy and Davis 1977).

To understand the role of the cerebellum in the control of stepping and scratching movements it is desirable to know how the principal mechanisms of these movements, i.e., the spinal ones, are organized. Unfortunately, so far their organization is unclear, and we shall consider only a few of the hypotheses concerning the arrangement of the spinal rhythmical generator.

The first attempt to explain a rhythmical process originating within the spinal cord was made by Brown (1911, 1914) who advanced a hypothesis on the inhibitory interaction between two "half-centres". According to the hypothesis in its modern form (Lundberg, 1969, 1981), the spinal cord contains two groups of neurons inhibiting each other. The inhibitory interaction is so strong that when one group (half-centre) is excited, activity of the other group is completely suppressed. In addition, there are excitatory connections within each half-centre, due to which all the cells are recruited into activity when some of them become active. To explain alternating activity of the half-centres, a process similar to fatigue is suggested. One half-centre is thought to exert excitatory influences on flexor motoneurons, while the other one acts on extensor motoneurons. Therefore, when a system of two half-centres generates a rhythmical process, flexors and extensors are activated alternately, a phenomenon that is observed (as a rough approximation) in stepping and scratching. A deficiency of this hypothesis is that transition of the activity from one half-centre to another is determined by "fatigue", i.e., a property which so far has not been found in spinal interneurons.

Another model claiming to explain the mechanism generating the rhythmical pattern underlying locomotion is that of a "ring-generator". The generator consists of a ring circuit of neurons successively exciting one another. Excitation propagates along the ring, the duration of one turn being equal to the cycle duration (Gurfinkel and Shik 1973; Gurfinkel et al. 1973). But so far, there are no experimental data on the existence of long neuronal chains in the spinal cord.

Finally, we shall consider a model with a "switch off mechanism". Several versions of this model were developed to explain rhythmical generation in the respiratory centre (Bradley et al. 1975; Wyman 1976; Cohen and Feldman 1977; Euler 1977, 1983), in the spinal cord during scratching (Berkinblit et al. 1978a) and stepping (Berkinblit et al. 1978b), as well as in the nervous system of the mollusk *Tritonia* during siwmming (Lennard et al. 1980). Here we shall consider the version of such a model developed to explain rhythmical generation during scratching. The network consists of two groups of neurons (Fig. 12C). There is one group of neurons with mutual excitatory connections (E-neurons). A part of them also exerts an excitatory action on a group of inhibitory neurons (I-neurons). The latter ones, in turn, strongly inhibit E-neurons. The whole system is in a state of stable equilibrium and has low activity since any increase of E-neuron activity is accompanied by a large increase of the inhibitory inflow to E-neuron from I-neurons. The system can be transferred into the state of rhythmical oscillations by inhibiting I-neurons ("switching on", Fig. 12C). E-neurons are thus released from inhibition, and their mutual excitation results in a regenerative process: new neurons are continuously recruited and discharge frequencies increase. When the activity of E-neurons is high enough the excitatory inflow to I-neurons reaches their thresholds, and I-neurons begin to fire. While firing, I-neurons inhibit E-neurons and, therefore, eliminate the source of their own excitation ("switch-off mechanism" comes into operation). The firing of both E- and I-neurons stops, and the system returns to its original state. Then a new cycle begins.

One can suppose that the rhythmical generator controls moto-neurons in the following way (Fig. 12C). E-neurons which are active by the end of the regenerative process excite extensor motoneurons. Flexor motoneurons are tonically activated during the switching on of the system (postural stage of the scratch reflex); then they are periodi-cally inhibited by I-neurons.

We find that this model fits the experimental data well. In partic-ular, it agrees with the data concerning the behaviour of spinal inter-neurons from "leading" segments during scratching described in the previous section. According to this model, the amount of simultaneously active units varies within a wide range during the cycle. Exactly the same phenomenon is observed during fictitious scratching (Fig. 12A) as well as during locomotion (Fig. 12B). The characteristic feature of group 2 (Fig. 11) is a gradual increase of the activity during the cycle: both the amount of simultaneously active neurons and their discharge frequencies increase (*freq* column in Fig. 11). One can assume that the neurons from group 2 as well as some neurons from group 3 correspond to E-neurons in the model, i.e., these units mutually excite each other, which brings about the increase of their total activity in the course of the cycle. Some of the group 3 neurons seem to be inhibitory ones (I-neurons in the model), since after their short-term excitation almost any activity of the spinal neurons disappears. It should be emphasized that the group 3 neurons are spontaneously active (*rest* column in Fig. 11). One can suppose that spontaneously active inhibitory neurons prevent an accidental onset of the regenerative process in the system and, therefore, contribute to its stability. During the latent period of rhythmical generation these units are inhibited (*stim* column in Fig. 11), which releases periodical oscillations in the system.

While studying spinal interneurons, a group of cells was distinguished (group 1 in Fig. 11) which became tonically active during elicitation of scratching. In the scratching cycle, the group 1 neurons fired during the L-phase and were inhibited in the S-phase. Apparently, these neurons do not participate in rhythmical generation; rather, they are driven by the generator. The pattern of the group 1 neuron activity is similar to that of flexor motoneurons. It is possible that group 1 mediates influences of the rhythmical generator upon flexor motoneurons.

There are some data in favour of the suggestion that rhythmical processes underlying stepping and scratching movements are generated by a common spinal mechanism. It was mentioned above that the rostral segments of the spinal hindlimb centre play a crucial role in generating both movements. Besides, in some cases, the scratch reflex began with long cycles, the flexor and extensor phases being the same which is typical for stepping; then a gradual transition from these long cycles to normal scratching cycles, with a very short extensor phase, could be observed (Berkinblit et al. 1978b). During this transition, spinal interneurons from the L4 and L5 segments preserved their phases of activity in the cycle. Finally, the phase distribution of spinal interneurons during stepping (Orlovsky and Feldman 1972b) has an important peculiarity which has also been emphasized while describing the phase distribution of neurons during scratching: in both cases the amount of active neurons increases in the flexor phase and decreases in the extensor phase (Fig. 12A, B). These experimental findings suggest that there is, in the rostral segments of the lumbo-sacral enlargement, a neuronal mechanism capable of generating rhythmical oscillations of various cycle durations and various lengths of the flexor and extensor parts of the cycle. The mode of activity of this mechanism is determined by descending commands: in pinna stimulation the scratching rhythm arises, in stimulation of the locomotor region the stepping one. Both modes of activity could be based on the principle of the generator with a "switch-off mechanism".

# II  Signals Coming to the Cerebellum

## 1.  Cerebellar Cortex and its Afferent Connections

The cerebellum consists of the cortex, the white matter and the three pairs of nuclei: the fastigial (FN), interpositius (IN) and lateral (LN) (Fig. 13A). Three longitudinal zones are distinguished in the cerebellar cortex: the vermis, the pars intermedia (paravermal zone of the hemi-spheres) and the lateral zone of the hemispheres, each of them sending fibres to one of the cerebellar nuclei. The vermis projects to the FN and to the lateral vestibular nucleus of Deiters; the pars intermedia projects to the IN; and the lateral zone of the hemisphere to the LN. In turn, the cerebellar nuclei project to various structures of the brain stem. Transversal sulci divide the cerebellar cortex into lobes and lobules (Fig. 13A).

The description of the cerebellar cortex and the interactions between cerebellar neurons may be found in a number of books and review

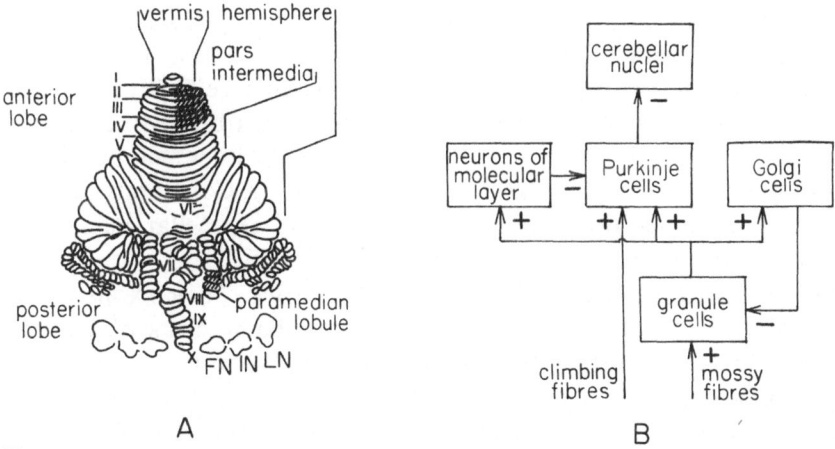

A                                                    B

**Fig. 13A, B.** The cerebellar cortex. *Roman numerals* show the lobules according to Larsell (1937, 1967). Hindlimb areas in the anterior lobe and paramedian lobule are *hatched*. *IN, FN* and *LN* — interpositus, fastigial and lateral nuclei. **B** Interaction between neurons in the cerebellar cortex (*plus,* excitatory synapse; *minus,* inhi-bitory one)

articles, which also provide full bibliographies (Ramón y Cajal 1911; Jansen and Brodal 1954; Fox 1962; Smolyaninov 1966; Eccles et al. 1967; Fox and Snider [eds] 1967; Larsell 1967; Llinás [ed] 1969; Palkovits et al. 1971a, b, c, 1972; Palay and Chan-Palay 1974; Armstrong 1978).

Cortical afferent inputs and interconnections of cortical neurons are schematically shown in Fig. 13B. There are two types of afferent fibres coming to the cerebellar cortex — mossy and climbing fibres[2]. The majority of the cerebellar inputs terminate as mossy fibres, those coming from the spinal cord in particular. Mossy fibres exert excitatory action upon granule cells of the cerebellar cortex which, in turn, excite Purkinje cells, neurons of the molecular layer and Golgi cells (Eccles et al. 1966a, b, c, 1967). Neurons of the molecular layer (basket and stellate cells) exert strong inhibitory action upon Purkinje cells (Andersen et al. 1964; Eccles et al. 1966a, 1967). Golgi cells inhibit granule cells (Eccles et al. 1966c, 1967).

The pathways formed by mossy fibres are characterized by extensive convergence and divergence of connections. A single mossy fibre terminates on 200–800 granule cells, the axon of a single granule cell terminates on 30–45 Purkinje cells; 3–5 mossy fibres converge on a single granule cell, and 20,000–30,000 axons of granule cells converge on a single Purkinje cell (Fox 1962; Smolyaninov 1966; Palkovits et al. 1971a, b, c, 1972).

Climbing fibres terminate directly on Purkinje cells, each cell obtaining one climbing fibre. Climbing fibres originate mainly from the inferior olives of the medulla (Szenthagothai, Rajkovits 1959). These fibres exert very strong excitatory action upon Purkinje cells (Eccles et al. 1966d, 1967). An impulse arriving via a climbing fibre evokes a high frequency (300–500 pulses $s^{-1}$) discharge of a Purkinje cell consisting of 4–6 spikes ("complex spikes", see Thach 1967). Collaterals of both mossy and climbing fibres terminate also in the cerebellar nuclei and in the nucleus of Deiters.

Purkinje cells are the "output" cortical neurons. Their axons terminate in the cerebellar nuclei and in the vestibular nucleus of Deiters. Purkinje cells are inhibitory neurons. Their stimulation results in monosynaptic inhibition of the neurons of cerebellar nuclei as well as of Deiters' nucleus (Ito et al. 1964b, 1966, 1968a, b, 1970c; Ito and Yoshida 1964; Ito 1965, 1967; Eccles et al. 1967).

---

[2] Recently, one more type of fibres coming to the cerebellar cortex was found, i.e., thin noradrenergic fibres, originating from the locus coeruleus of the brain stem (Hökfelt and Fuxe 1969; Bloom et al. 1971; Olson and Fuxe, 1971). However, the data on the pathways terminating as noradrenergic fibres are sparce so far

Signals from the somatic receptors reach two areas of the cerebellum: lobules II–VI in its rostral part and VII–VIII in the caudal part (Fig. 13A). Electrophysiological studies have revealed somatotopic organization of the somatosensory cerebellar inputs (Adrian 1943; Snider and Stowell 1944; Snider 1952; Dow and Moruzzi 1958; Kitai et al. 1969; Eccles 1970; Arshavsky 1972). Stimulation of the hindlimb nerves or receptors causes responses mainly in the rostral part of the anterior lobe (lobules II–IV), both in the vermis and in the pars intermedia, as well as in the caudal part of the paramedian lobule. These areas are hatched in Fig. 13A. Stimulation of the forelimb nerves or receptors causes responses in the caudal part of the anterior lobe (lobule V) and in the rostral part of the paramedian lobule. Finally, stimulation of the receptors of the head causes responses in lobule VI and in the very rostral folium of the paramedian lobule. Somatosensory projections manifest much more detailed somatotopic specificity in the pars intermedia than in the vermis.

Similar somatotopic organization was observed while studying the cerebellar efferent influences on motor activity. Stimulation of the fore- or hindlimb areas of the cerebellar cortex results in motor responses of the corresponding limb, stimulation of the pars intermedia evoking more local motor responses than that of the vermis (Hare et al. 1937; Hampson et al. 1952; Pompeiano 1967).

However, it should be mentioned that the cerebellar cortex is characterized by rather poor topographical discrimination. Even in the pars intermedia, where the somatotopic organization of afferent projections is manifested better than in the vermis, Purkinje cells produce similar responses to stimulation of nerves or receptors of different muscles (including antagonistic ones) of the given limb (Eccles et al. 1971b; Murphy et al. 1973). It follows that in the cerebellar cortex, in contrast to the cerebral motor cortex (see Asanuma 1975), separate representation of individual muscles is absent.

Signals from the spinal hindlimb centres reach the cerebellum via four principal pathways: dorsal spino-cerebellar tract (DSCT), ventral spino-cerebellar tract (VSCT), spino-reticulo-cerebellar pathway (SRCP), and spino-olivo-cerebellar pathway (SOCP). To study signals arriving via these pathways during locomotion and scratching, activity of the DSCT and VSCT neurons as well as of reticulo-cerebellar and olivo-cerebellar neurons was recorded.

## 2. Dorsal Spino-Cerebellar Tract

### a) General Characteristics

The DSCT consists of thick, fast conducting $(30-110$ m $s^{-1})$ fibres (Oscarsson 1965, 1967, 1973). The cell bodies of the DSCT neurons lie in Clarke's column. The DSCT ascends in the dorsal part of the ipsilateral lateral funiculus and enters the cerebellum through the restiform body.

The DSCT terminates as mossy fibres in the pars intermedia and in the adjacent lateral vermis of the rostral part (lobules II—IV) of the anterior lobe as well as in the caudal part of the paramedian lobule, i.e., in the hindlimb areas of the cerebellar cortex (Fig. 13A). The branching of the DSCT mossy fibres in the cerebellar cortex is rather narrow. Antidromic response of the DSCT neurons can be evoked from the cortex surface of approx. 1 mm$^2$ (Lundberg and Oscarsson 1960). On the way to the cerebellar cortex, the DSCT fibres give rise to collaterals terminating in Deiter's nucleus (see Chap. III, Sec. 1) and, probably, to a smaller extent, in the cerebellar nuclei (Matsushita and Ikeda 1970b).

Afferent connections of DSCT neurons have been thoroughly studied (Lloyd and McIntyre 1950; Laporte et al. 1956a, b; Lundberg and Oscarsson 1956, 1960; Lundberg and Winsbury 1960; McIntyre and Mark 1960; Eccles et al. 1961b; Jansen and Rudjord 1965; Oscarsson 1965, 1967, 1973; Kostyuk 1969; Vasilenko et al. 1969; Jansen and Walloë 1970; Lindström and Takata 1972; Mann 1973). The main input to DSCT neurons is formed by the afferent fibres of group Ia coming from the primary receptors of muscle spindles and of group Ib from the Golgi tendon organs; the muscle spindle and tendon organ afferents activating different DSCT neurons. The DSCT neurons can also receive monosynaptic influences from the secondary receptors of the muscle spindles, from the joint and foot pad receptors, as well as polysynaptic influences from the flexor reflex afferents.

The efficiency of transmission from group I afferents to the DSCT neurons is high. The unitary EPSP's evoked by these afferents are up to 5 mV in amplitude (Eccles et al. 1961b; Kuno and Miyahara 1968; Eide et al. 1969a, b). Thus, the generation of action potentials does not require wide spatial summation of afferent impulses (Lloyd and McIntyre 1950; Holmqvist et al. 1956; McIntyre and Mark 1960; Kostyuk 1969; Vasilenko et al. 1969). Such an efficient synaptic linkage seems to be accounted for by the "giant" synapses formed by primary afferents on the DSCT neurons; the size of end bulbs being up to several hundred square microns (Szenthagothai and Albert 1955).

The DSCT neurons can follow the frequencies of input signals up to several hundred pulses per second (Holmqvist et al. 1956; McIntyre and Mark 1960; Jansen and Rudjord 1965; Jansen et al. 1966, 1967b; Pyatigorsky 1968; Kostyuk 1969; Jansen and Walloe 1970). The discharge frequency of the DSCT neurons is linearly related to muscle length, varying within a considerably wide range. The ability to follow high frequency input signals can be explained by (1) high efficiency of the synaptic transmission; (2) short duration of the action potentials; (3) low level of after-hyperpolarization; and (4) the lack of recurrent inhibition in the DSCT neurons (Eccles et al. 1961b; Kuno and Miyahara 1968; Eide et al. 1969a, b).

Receptive fields of the DSCT neurons are rather small. The excitatory actions of group I afferents usually arise from a single muscle or a few synergists of the ipsilateral hindlimb. Besides, activation of the stretch receptors of some other muscles, both antagonists and synergists, results in the inhibition of the DSCT neurons (Curtis et al. 1958; Eccles et al. 1961b, 1963; Oscarsson 1965, 1967, 1973; Jansen et al. 1967a; Vasilenko et al. 1969; Hongo et al. 1983a, b).

Even if supraspinal influences on the DSCT neurons exist, they are very small, at least in comparison to the influences upon neurons of other spino-cerebellar tracts described below (Lundberg et al. 1963; Hongo and Okada 1967; Hongo et al. 1967; Kostyuk 1969; Vasilenko et al. 1969; Kubota and Poppele 1977).

The present data allowed Oscarsson to conclude that via the DSCT the intermediate area of the cerebellar cortex receives information on the activity of single muscles or small groups of synergists (Oscarsson 1965, 1967, 1973).

b) Activity of DSCT Neurons During Locomotion

The activity of DSCT neurons was studied during locomotion in mesencephalic cats evoked by stimulation of the locomotor region (Arshavsky et al. 1972a, d 1974). The DSCT neurons were identified by their antidromic response to stimulation of the hindlimb area in the pars intermedia of the cerebellar anterior lobe (Fig. 14 A–E).

Stimulation of the locomotor region has practically no influence on the activity of DSCT neurons during the period preceding the beginning of locomotion. When locomotion begins, activity of the DSCT neurons becomes rhytmical and is strictly linked with stepping movements: the

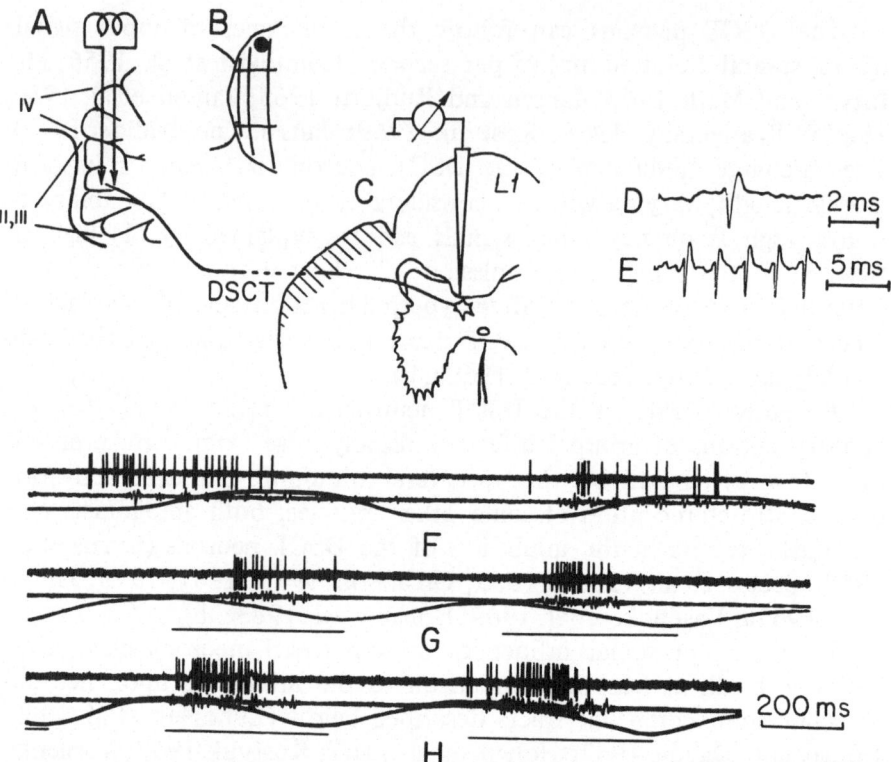

**Fig. 14 A—H.** Recording and identification of DSCT neurons. Neurons were recorded from Clark's column in the *L1* segment and identified by the antidromic response to stimulation of the hindlimb area in the pars intermedia of the cerebellar anterior lobe. The site of insertion of the stimulating electrode is shown in **B** (*black dot*), its position in the cortex (in the *lobules II, III*) is shown in **A**. Position of the recording electrode near a DSCT neuron and location of the DSCT in the spinal cord (*hatched area*) are shown in **C**. **D, E** Antidromic responses of a neuron to a single stimulus and to a train of impulses (370 pps) applied to the cerebellum. **F—H** Activity of a DSCT neuron that is excited by afferents from ankle extensors. **F** Response of the neuron to passive flexion of the ipsilateral ankle joint. **G, H** Activity of the neuron during weak (**G**) and intense (**H**) locomotion. In **F—H**, besides the activity of the DSCT neuron, EMG of m.gastrocnemius and the angles at the ankle joint (in **F**) or at the hip joint (in **G, H**) are shown. The upward deflection of the trace corresponds to flexion. *Horizontal lines* in **G** and **H** indicate the stance phases of the ipsilateral hindlimb (Arshavsky et al. 1972a, d)

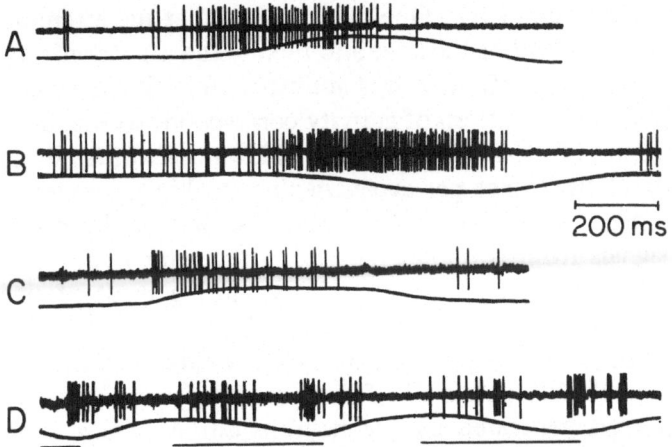

**Fig. 15 A—D.** Activity of a DSCT neuron that is excited by afferents from two-joint muscles. **A—C** Responses of the neuron to the knee flexion (**A**), to the hip extension (**B**), and to the pressure applied to muscles on the fore surface of the femur (**C**). **D** Activity of the neuron during locomotion. The *lower traces* are the knee angle (in **A**; flexion—up), the hip angle (in **B**, **D**; flexion—up) or the period of the muscle pressure (in **C**); *horizontal lines* in **D** indicate the stance phases of the ipsilateral hindlimb (Arshavsky et al. 1972a, d)

cells fire in bursts separated by periods of silence (Figs. 14 and 15)[3]. The comparison of the behaviour of DSCT neurons during passive movements of the ipsilateral hindlimb at different joints and during locomotion has shown that DSCT neurons fire during the active contraction of the muscles (or muscle) sending afferents to the given neuron. Fig. 14 shows a DSCT neuron which was activated by passive flexion of the hindlimb at the ankle joint simultaneously with reflex excitation of m.gastrocnemius (F). This neuron did not react to limb movements at any other joint. Thus, one may conclude that afferents from the ankle extensors terminate on this neuron. During locomotion, the neuron periodically fired in the stance phase of the step, i.e., during active contraction of m.gastrocnemius (G, H). As stimulation of the locomotor region got stronger, the extensor activity increased, limb movements became more forceful, and the activity of the neuron also markedly increased (cf. G and H).

Figure 15 shows another neuron, which was activated by both knee flexion (A) and hip extension (B) as well as by palpation of muscles on

---

[3] We shall use the term "rhythmical activity" to designate the activity linked with the rhythmical process generated by the spinal cord

the rostral surface of the femur (C). One may suggest that the afferents from two-joint muscles, extending the knee and flexing the hip, terminate on this neuron. During locomotion, it was activated in both the stance and swing phases (D). Such a pattern of activity corresponds to the more complicated behaviour of two-joint muscles (like m.sartorius, 8 in Fig. 4), which contract two times per cycle, flexing the hip and extending the knee. A similar pattern of activity was observed in the other neurons receiving signals from two-joint muscles. They generated either two bursts of impulses per step or a long burst during most of the step.

The maximum discharge frequencies of DSCT neurons during in tensive locomotion reach 200 pulses $s^{-1}$. These frequencies are comparable to the highest ones generated by the neurons during passive limb movements.

Figure 16A shows the phase distribution of DSCT neurons. As in Figure 11, the duration of the locomotor cycle is assumed to be a unit, the interrupted line shows the average time of transition between the swing and stance phases. Horizontal lines show burst positions of DSCT neurons in the locomotor cycle. One can see that the DSCT neurons are active either in the stance phase (as the cell in Fig. 14) or both in the stance and swing phases (as the cell in Fig. 15). None of the recorded neurons was active in the swing phase only. This might be accounted for by the fact that "pure" flexors represent only a small part of the hindlimb muscles (Reighard and Jennings 1935).

Figure 16B shows the activity of the "average" DSCT neuron plotted as a function of the phase of the step (the "frequency curve"). To obtain the curve, the locomotor cycle was divided into seven intervals (three of them in the swing phase and four in the stance phase),

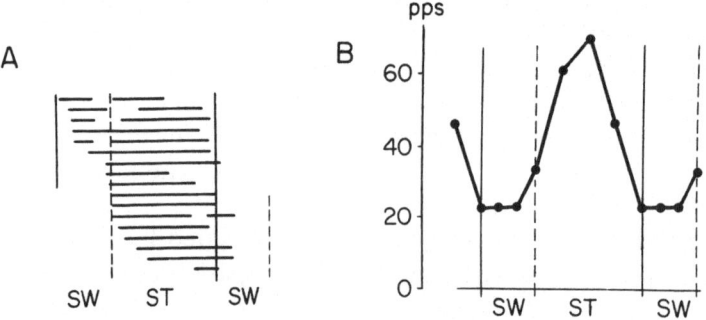

**Fig. 16A, B.** Relationship between the activity of DSCT neurons and the phase of the step of the ipsilateral hindlimb. **A** The phase distribution of 17 DSCT neurons. The burst position of each neuron in the normalized step cycle is shown by a *horizontal line.* **B** The discharge frequency (pulses per second) of the "average" DSCT neuron as a function of the phase of the step. The swing (*SW*) and stance (*ST*) phases of the step cycle are indicated (Arshavsky et al. 1972d)

and the discharge frequency of the neurons was summed for each interval. Then the average frequency for all the neurons in the given interval was calculated. Since the neurons were recorded in different experiments with different intensity of locomotion, the curve presents only a rough characteristic of the average activity of the population of DSCT neurons. Still to a certain extent it reflects the overall flow of impulses in the DSCT in different phases of the step. This flow sharply increases during the first half of the stance phase, reaching the maximum about the middle of the phase, and then abruptly decreases. Such a pattern of DSCT activity may be explained by the fact that at the beginning of the stance phase most of the hindlimb muscles (extensors and the majority of two-joint muscles) are activated; they support the body weight and provide the propulsive force. Thus, the total afferent inflow from muscle receptors must also reach the maximum in this phase of the step. Besides, those DSCT units which receive inputs from receptors located in the foot pads must also be activated in the stance phase.

Thus, the comparison of the data described here and in Sec. 5, Chap. I suggests that the pattern of DSCT activity during locomotion resembles the pattern of activity of the muscle receptors of the hindlimb (primary spindle endings and Golgi organs). Both muscle receptors (see Fig. 8B, C) and DSCT neurons are active mainly during contraction of the corresponding muscle. This similarity suggests that rhythmical activity of the DSCT neurons is determined mainly by the afferent signals, namely, by those from muscle receptors and, to some extent, from pad and joint receptors. The crucial role of afferent signals in generating this activity was proved through recording DSCT neurons during locomotion in cats with deafferented hindlimbs, as well as during fictitious scratching. In both cases, any rhythmical modulation in DSCT neurons was absent (Fig. 17A, B).

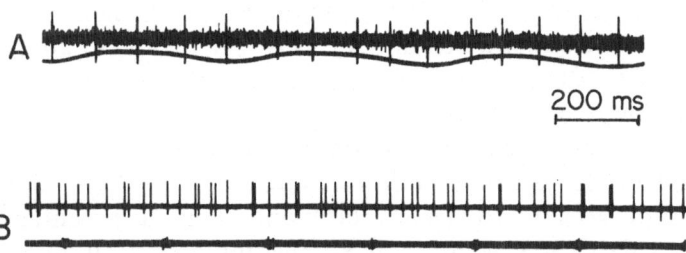

**Fig. 17A, B.** Activity of DSCT neurons during locomotion of the cat with deafferented hindlimbs (**A**) (Arshavsky et al. 1972d) and during fictitious scratch reflex (**B**) (Arshavsky and Pavlova, unpublished). The *lower trace* is the hip angle of the ipsilateral hindlimb (in **A**) or ENG of n.gastrocnemius (in **B**)

## 3. Ventral Spino-Cerebellar Tract

### a) General Characteristics

The VSCT contains thick fibres; the conduction velocity being 40–140 m s$^{-1}$ (Lundberg and Oscarsson 1962a; Oscarsson 1965, 1967, 1973; Burke et al. 1971). VSCT neurons are located in the L3–L6 segments of the spinal cord. Two sources of VSCT fibres are known: (1) Cooper-Sherrington cells, located in the ventrolateral border of the ventral horn (Cooper and Sherrington 1940; Burke et al. 1971) and (2) the cells located in the intermediate zone of the grey matter, i.e., in the lateral part of Rexed's layers V–VII (Hubbard and Oscarsson 1962). The majority of VSCT fibres crosses the midline, ascends in the ventral part of the contralateral lateral column and, after entering the cerebellum, crosses the midline once again. The VSCT terminates as mossy fibres in the cerebellar anterior lobe, in the vermis and pars intermedia (lobules II–IV) (Grant 1962; Lundberg and Oscarsson 1962a; Burke et al. 1971), i.e., in the hindlimb area (the upper hatched area in Fig. 13A). In contrast to the DSCT, the VSCT does not terminate in the cerebellar posterior lobe.

The VSCT fibres branch extensively in the cerebellar cortex and innervate relatively large cortical areas. Antidromic responses of VSCT neurons can be evoked from different points within a cortical area of as much as 25 mm$^2$, which is about 25 times larger than the area found for DSCT neurons; some of the VSCT fibres have bilateral terminations (Lundberg and Oscarsson 1962a; Burke et al. 1971; Arshavsky et al. 1978b). On their way to the cortex, VSCT fibres give out collaterals, terminating in the cerebellar nuclei (see Chap. V).

Electrophysiological experiments have shown that many VSCT neurons are activated by group I muscle afferents (Oscarsson 1957, 1965, 1967, 1973; Eccles et al. 1961a; Lundberg and Weight 1971; Lindström 1973; Lindström and Schomburg 1974). Afferents of groups Ia and Ib terminate on different VSCT neurons. The group Ia afferents terminate mainly on the neurons located in the border area of the ventral horn, while the group Ib afferents terminate on the neurons located in the lateral part of the layers V–VII. The VSCT neurons have wide receptive fields. Monosynaptic EPSPs (excitatory postsynaptic potentials) may be evoked in the VSCT neurons by stimulation of several muscle nerves, especially of those innervating muscles of various joints. Correspondingly, VSCT neurons usually respond to passive movements in different joints of the ipsilateral hindlimb and, to a smaller extent, of the contralateral one (Fig. 18 G–J). The efficiency of excitatory influence of muscle afferents upon VSCT neurons is considerably lower than upon

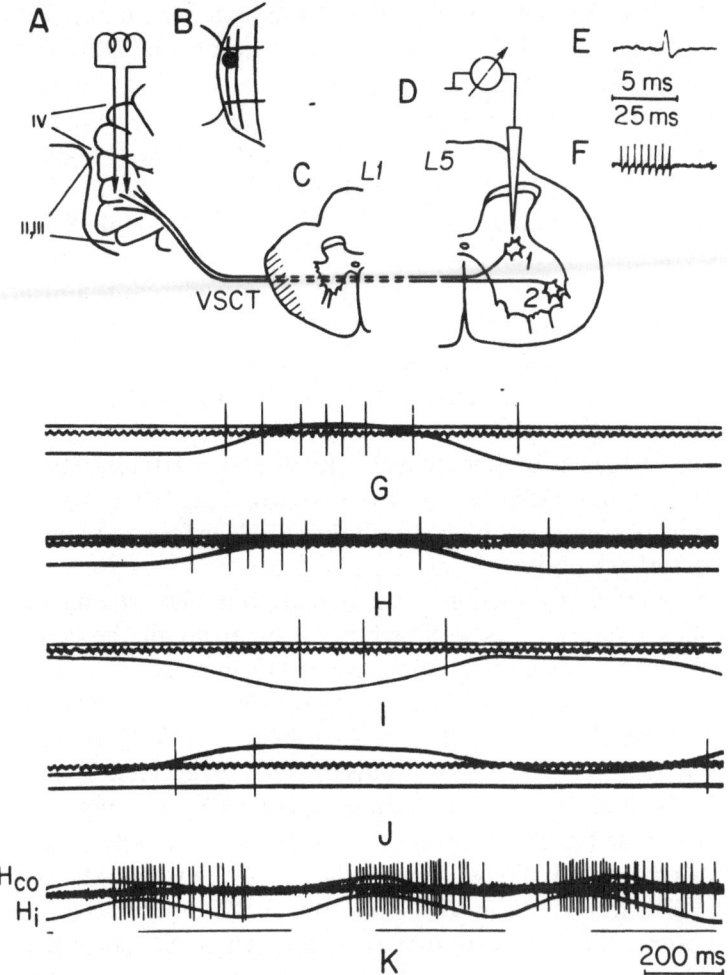

**Fig. 18 A–D.** Recording and identification of VSCT neurons. Neurons were recorded from the intermediate zone of the grey matter (*1*) or from the border zone of the ventral horn (*2*) in the *L5* (or in *L4*) segment (**D**), and identified by the antidromic response to stimulation of the hindlimb area in the vermis (marked in **B**) or in the pars intermedia of the anterior lobe. Location of the VSCT in the contralateral part of the spinal cord at the L1 level (*hatched area*) is shown in **C**. **E, F** Antidromic responses of a neuron to a single stimulus and to a train of impulses (400 pps) applied to the cerebellum. **G–J** Responses of a VSCT neuron to passive limb movements. **G–I** Movements of the ipsilateral hindlimb at the ankle (**G**), knee (**H**) and hip (**I**) joints (flexion–up). **J** Movement of the contralateral hindlimb at the hip joint (extension–up). **K** Activity of the same neuron during locomotion. Movements are also shown at the ipsilateral and contralateral hip joints ($H_i$, flexion–up; $H_{co}$, extension–up) as well as the stance phases of the ipsilateral hindlimb (*horizontal lines*) (Arshavsky et al. 1972e)

DSCT neurons. This can be clearly seen from comparison of Figs. 14F and 15A, B with Fig. 18 G—J, which show reactions of DSCT and VSCT neurons to passive limb movements.

Stimulation of the group I afferents results also in di- and poly-synaptic IPSP's (inhibitory postsynaptic potentials) in VSCT neurons. The convergence pattern of excitatory and inhibitory inputs to VSCT neurons is highly complicated. According to Lundberg and Weight (1971), if all convergent excitatory and inhibitory effects from primary afferents as well as from descending pathways are considered, it appears that one VSCT neuron hardly resembles another.

VSCT neurons also receive strong polysynaptic inputs, both exitatory and inhibitory, from the flexor reflex afferents. For a part of the VSCT neurons, the flexor reflex afferents are the sole source of excitatory influence. The convergence of signals from flexor reflex afferents to the VSCT neurons is very extensive. Many of the neurons respond to stimulation of cutaneous afferents and high threshold muscle afferents from the whole caudal part of the body.

VSCT neurons receive not only peripheral, but also strong supraspinal influences, excitatory as well as inhibitory, from all the central structures giving rise to descending spinal tracts (Holmqvist et al. 1960; Oscarsson 1960, 1965, 1973; Eccles et al. 1961a; Magni and Oscarsson 1961; Lundberg and Oscarsson 1962a; Lundberg and Weight 1971; Baldissera and Roberts 1975, 1976; Baldissera and Bruggencate 1976; Fu et al. 1977). Stimulation of the vestibulo-spinal (VS), reticulo-spinal (RS) and rubro-spinal (RbS) tracts results in mono- and polysynaptic EPSP's and polysynaptic IPSP's while stimulation of the pyramidal tract polysynaptic EPSP's and IPSP's in VSCT neurons. VSCT neurons respond also to natural stimulation of vestibular receptors when the cat is tilted in the frontal plane (Arshavsky et al. 1972e, f). The tilt evokes reactions of neurons of the vestibulo-spinal and other descending tracts (Orlovsky and Pavlova 1972a; Panchin 1978. Finally, the transmission from the flexor reflex afferents to VSCT neurons (and to other neurons as well) is faciliated by the pyramidal tract and inhibited by the reticulo-spinal tract.

Lundberg (1959, 1964, 1966, 1971) and Oscarsson (1967, 1969a, 1973) suggested that some of the spino-cerebellar pathways (including the VSCT), which originate from neurons receiving peripheral signals mainly from the flexor reflex afferents, have wide receptive fields and are strongly influenced by supraspinal centres, are not sensory pathways in the common sense. They convey information not on peripheral events (limb movements, muscle contractions, etc.) but, rather, on the activity of central spinal mechanisms. This hypothesis was given detailed consideration by Lundberg (1971) in application to the VSCT. Lundberg suggested that the VSCT carries information on transmission in inhibitory reflex pathways. A number of experiments carried out in Lundberg's laboratory has shown that an inhibition of VSCT neurons evoked by stimulation of different somatic nerves and supraspinal structures is mediated by the axon collaterals of those interneurons which inhibit motoneurons (Gustafsson and Lindström 1973; Lindström 1973; Lindström and Schomburg 1973, 1974; Baldissera and Roberts 1975, 1976; Baldissera and Bruggencate 1976; Fu et al. 1977). According to these authors, their findings prove the validity of Lundberg's hypothesis that the VSCT monitors the transmission in the inhibitory pathways of the spinal cord. We shall consider this hypothesis later, discussing the results obtained while studying the VSCT neuron activity during locomotion and scratching.

b) Activity of VSCT Neurons During Locomotion

The activity of VSCT neurons, was studied during locomotion of mesencephalic cats (Arshavsky et al. 1972b, e, f, 1974). The neurons were recorded from the L4 and L5 segments of the spinal cord, both from the intermediate area of the spinal grey matter and from the border area of the ventral horn (neurons 1 and 2 in Fig. 18D). Neurons were identified by the antidromic response to stimulation of the hindlimb zone in the vermis or in the pars intermedia of the cerebellar anterior lobe (Fig. 18 A–F). Stimulation of the "locomotor region" results in the increase of the discharge frequency (from 2–10 to 30–100 pulses $s^{-1}$) during the period preceding the beginning of locomotion in about half of the VSCT neurons. As locomotion starts, all the VSCT neurons begin to fire periodically with the rhythm of stepping (Fig. 18K). The units generate one burst per step. The more intensive locomotion is, the longer the duration of bursts and the higher the discharge frequency within the burst. During intense locomotion, the discharge frequency reaches 100–200 pulses $s^{-1}$.

Figure 19A shows the phase distribution of VSCT neurons in the locomotor cycle. One can see that the bursts of various VSCT neurons are distributed throughout the cycle, but the neurons active in the swing phase prevail. The frequence curve of the "average" VSCT neuron also shows that the overall flow of impulses in the VSCT increases during the swing phase of the step, reaching the maximum by the end of this phase, and then decreases during the stance phase (Fig. 19B).

The analysis of the behaviour of VSCT neurons suggests that their rhythmical activity, unlike that of DSCT neurons, is not determined by signals coming from the muscle receptors. Indeed, the discharge frequency of VSCT neurons during locomotion (100–200 pulses $s^{-1}$,

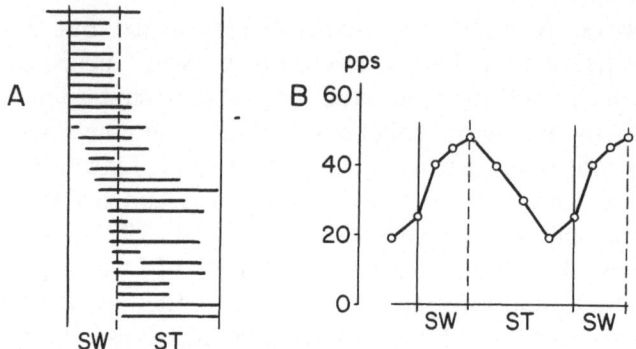

**Fig. 19A, B.** Relationship between the activity of VSCT neurons and the phase of the step of the ipsilateral hindlimb. **A** The phase distribution of 30 VSCT neurons. **B** The discharge frequency of the "average" VSCT neuron as a function of the phase of the step. Abbreviations as in Fig. 16 (Arshavsky et al. 1972e)

Fig. 18K) is much higher than during passive limb movements (10–20, maximum 50 pulses $s^{-1}$, Fig. 18 G–J). The overall activity of VSCT neurons is maximum in the swing phase of the step and minimum in the stance phase. However, as it has already been mentioned, the afferent inflow from muscle receptors is maximum in the stance phase. This suggests that the rhythmical activity of VSCT neurons is determined not by the signals coming from the receptors of the moving limb, but by other factors.

Further investigations of VSCT neurons were stimulated by Lundberg's idea about "intraspinal inputs" to these neurons. The activity of

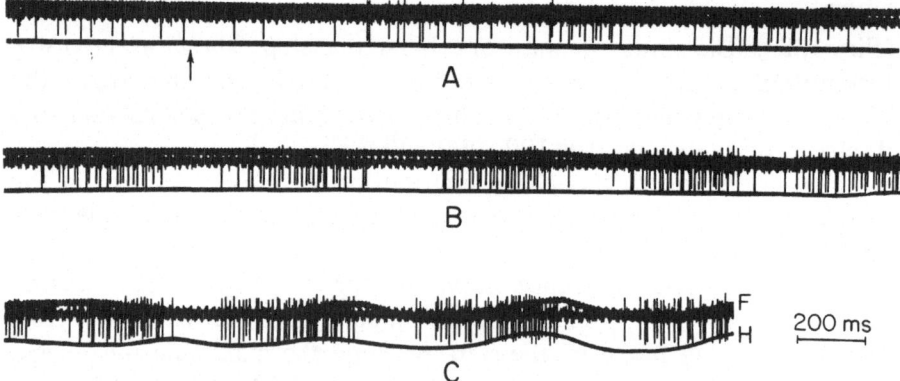

**Fig. 20 A–C.** Activity of a VSCT neuron during locomotion of a cat with deafferented hindlimbs. **A–C** is a continuous recording of the activity of the VSCT neuron and of movements of the ipsilateral fore (*F*) and contralateral hind (*H*) limbs (protraction of a limb corresponds to upward deflection of the beam). The *arrow* indicates the beginning of stimulation of the locomotor region (Arshavsky et al. 1972f)

VSCT neurons during locomotion in cats with deafferented hindlimbs was studied (Arshavsky et al. 1972b, f, 1974). It was found that the behaviour of VSCT neurons is practically the same both in animals with intact innervation and in those with deafferented hindlimbs. During locomotion the neurons fire periodically with the rhythm of stepping movements (Fig. 20), the discharge frequency being 50–100 (sometimes up to 150) pulses s$^{-1}$. In some cases, discharge modulation arises before any limb movements began (Fig. 20A, B).

Figure 21 shows the phase distribution of neurons (A) and the frequency curve for the "average" neuron (B) during locomotion in a cat with deafferented hindlimbs. One can see that, as in animals with intact innervation of the limbs (Fig. 19), the overall activity of neurons is increasing during the swing phase, reaching the maximum by the end of this phase, and decreasing during the stance phase.

What is the reason for rhythmical activity of VSCT neurons in animals with deafferented hindlimbs? Firstly, it should be noticed that stimulation of the "locomotor region" evokes stepping of the deafferented hindlimbs provided that the forelimbs also perform stepping movements (Shik et al. 1966a). On the other hand, VSCT neurons respond to passive movements of the forelimbs. Therefore, in cats with deafferented hindlimbs, the rhythmical activity of VSCT neurons might be determined by signals coming from the forelimbs. Had it been so, the period of this activity would have always coincided with the period of the forelimb movements. The actual situation turned out to be different. Figure 22A, B shows a case when periodical movements of the hindlimbs were spontaneously disturbed: within the interval marked by a horizontal line, the hindlimb stepped twice while the forelimb performed only one step. Correspondingly, the VSCT neuron fired twice, i.e., bursts were related with hindlimb movements. Another example is shown in Fig. 22C. Here, the fore- and hindlimbs were stepping with different

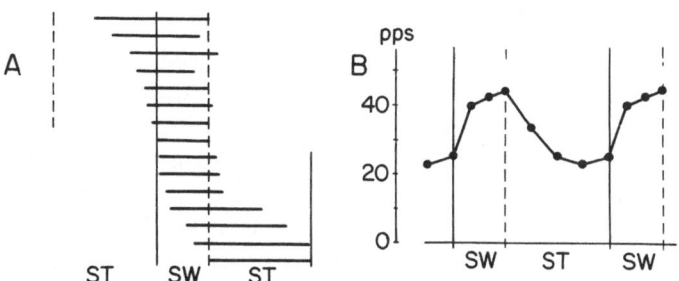

**Fig. 21A, B.** Relationship between the activity of VSCT neurons and the phase of the step of the ipsilateral hindlimb during locomotion of cats with deafferented hindlimbs. **A** The phase distribution of 15 VSCT neurons. **B** The discharge frequency of the "average" VSCT neuron as a function of the phase of the step. Abbreviations as in Fig. 16 (Arshavsky et al. 1972f)

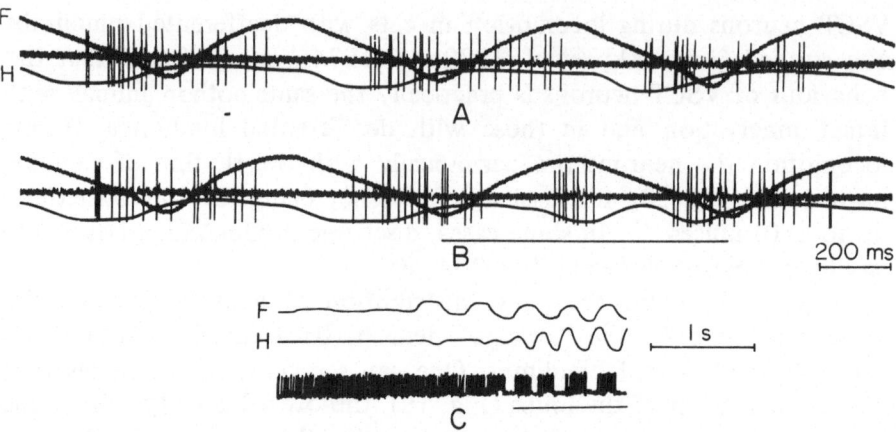

**Fig. 22 A–C.** Activity of two VSCT neurons (**A**, **B** and **C**) during locomotion of cats with deafferented hindlimbs when the fore (*F*) and hind (*H*) limbs are stepping with different rhythms. In **B**, two steps of the hindlimb during the period of one step of the forelimb are marked by a *horizontal line*. In **C** another example is shown: a *horizontal line* indicates 3 steps of the forelimb during 4 steps of the hindlimb (Arshavsky et al. 1972b, f)

rhythms. As a result, in the time interval marked by a horizontal line, the cat performed three steps by the forelimbs and four steps by the hindlimbs. One can see that also in this case firing of the neuron is related to movements of the hind-, but not of the forelimbs.

Thus, rhythmical activity of VSCT neurons can be observed in the absence of any corresponding rhythmical inflow from receptors of the executive apparatus. Corresponding results were obtained by Perret and colleagues (Perret et al. 1972; Perret 1976). During fictitious locomotion in decorticate curarized cats (see Chap. I), i.e., in the absence of rhythmical afferent signals from the limb receptors, they found rhythmically active fibres ascending in the ventral part of the lateral column. The authors suggest that at least some of these fibres belong to the VSCT.

It will be demonstrated in the next chapter that during locomotion all the descending brain stem-spinal tracts convey rhythmical signals. On the other hand, as was mentioned in Sect. 3 a, Chap. II, the descending tracts strongly affect VSCT neurons. Therefore, supraspinal centres might be a source of rhythmical activity of VSCT neurons. To test this hypothesis, the activity of VSCT neurons was studied in the decerebellate animals with deafferented hindlimbs, since any rhythmical signals in descending tracts are absent in such animals (see Chap. III). It was found that in this case the rhythmical activity of VSCT neurons is still preserved.

Thus, neither the signals from the moving limbs, nor the supraspinal influences are necessary for generating the rhythmical activity in VSCT neurons. Therefore, this activity (at least in cats with deafferented hindlimbs) is determined by the central spinal mechanism.

## c) Activity of VSCT Neurons During Scratching

The behaviour of VSCT neurons was studied mainly in immobilized thalamic cats, i.e., during fictitious scratching (Arshavsky et al. 1975a, 1978b). When scratching was evoked on the side ipsilateral to a neuron, VSCT neurons were rhythmically active, their bursts being related to rhythmical activity of motoneurons (Figs. 23; 26A, C; 27B, D). With more intense scratching, the rhythmical activity of the neurons increased (the bursts became longer, the discharge rate higher, cf. Fig. 23C and D). The discharge frequency during intense scratching reached 50–150 (up to 200) pulses s$^{-1}$.

Figure 24A shows the phase distribution of VSCT neurons during fictitious scratching. As in locomotion (Figs. 19A, 21A), the neurons are more active in the flexor (L) phase of the cycle than in the extensor (S) phase. The process of switching on of the neurons is distributed rather evenly throughout the cycle. The process of switching off is more abrupt: the majority of neurons stops firing at the end of the L-phase and in the S-phase. As a result, the number of simultaneously active neurons increased during the greater part of the L-phase and then rapidly

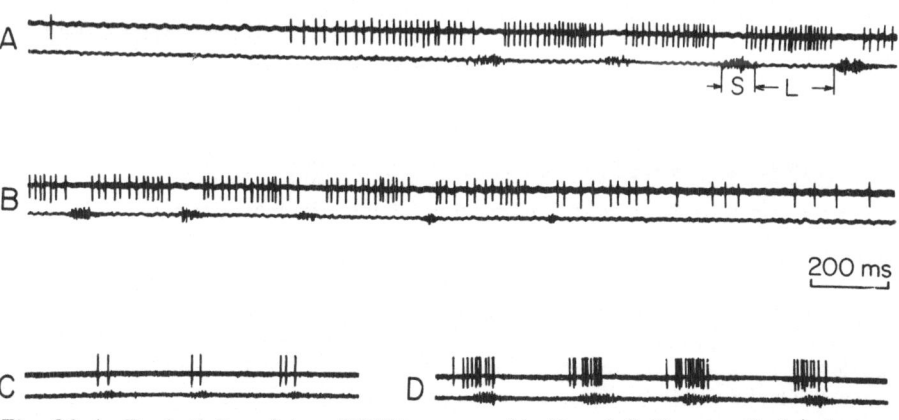

Fig. 23 A–D. Activity of two VSCT neurons (A, B and C, D respectively) during fictitious scratching. The pinna stimulation was started at the beginning of A and stopped at the beginning of B (B is the continuation of A). Activity of another neuron is shown during weak (C) and intense (D) scratching. The *lower trace* is the gastrocnemius ENG (Arshavsky et al. 1978b)

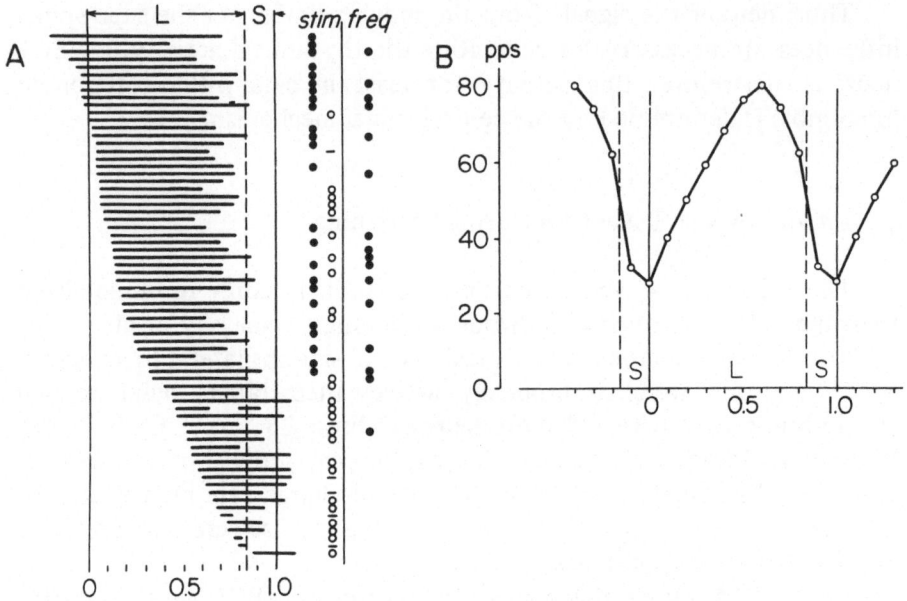

**Fig. 24A, B.** Relationship between the activity of VSCT neurons and the phase of the normalized scratch cycle. **A** The phase distribution of 69 VSCT neurons and the table of their parameters. In the column *stim*, the behaviour of most neurons during the latent period of scratching is shown (*filled circle*, facilitation; *open circle*, no change; *minus*, inhibition of the activity). In the column *freq*, the neurons firing at the rate increasing in the course of the burst are indicated. **B** The discharge frequency of the "average" VSCT neuron (pulses per second) as a function of the phase of the cycle (Arshavsky et al. 1978b)

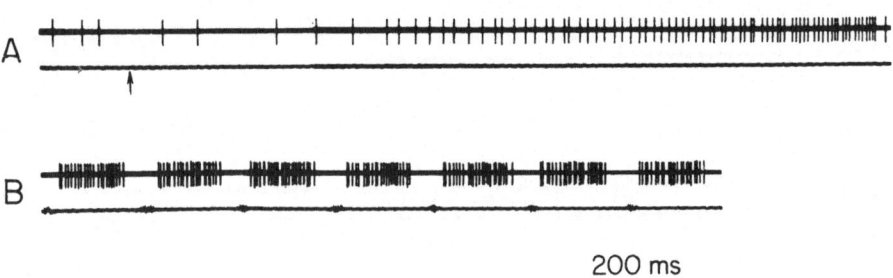

200 ms

**Fig. 25A, B.** Activity of a VSCT neuron during fictitious scratching in the decapitate cat (**B** is the continuation of **A**). The beginning of spinal cord stimulation (at the C1 level) is marked by the *arrow* in **A**. The *lower trace* is the gastrocnemius ENG (Arshavsky et al. 1978b)

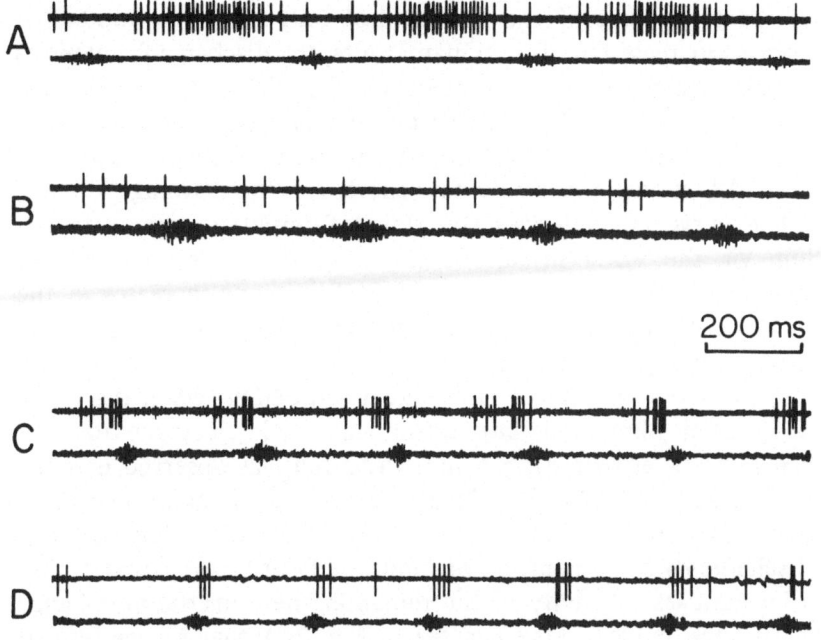

**Fig. 26 A–D.** Activity of two VSCT neurons (**A, B** and **C, D** respectively) during ipsilateral (**A, C**) and contralateral (**B, D**) fictitious scratching. The *lower trace* is the ENG of ipsilateral (**A, C**) or contralateral (**B, D**) n.gastrocnemius (Arshavsky et al. 1978b)

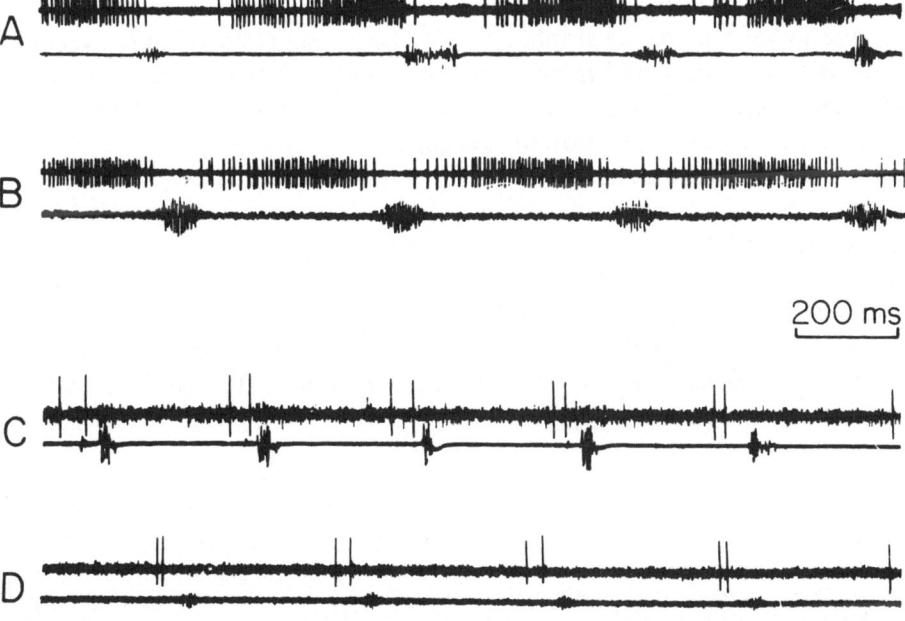

**Fig. 27 A–D.** Activity of two VSCT neurons (**A, B** and **C, D** respectively) recorded initially during actual scratching (**A, C**) and then (**B, D**) during fictitious scratching after Flaxedil injection (5 mg kg$^{-1}$, i.v.) The *lower trace* in **A, C** is the gastrocnemius EMG, and in **B, D** – the ENG (Arshavsky et al. 1978b)

decreased. In the course of the L-phase not only does the number of active neurons increase but also, in many units, the discharge frequencies become higher (sometimes, two-threefold, Fig. 26A, C; 27B). Such neurons are marked by filled circles in Fig. 24A (*freq* column). One can see that they are located in the upper and middle parts of the graph.

Figure 24B shows the frequency curve for the "average" VSCT neuron. The curve characterizes the flow of impluses coming to the cerebellum via the VSCT in various phases of the cycle. One can see that the activity of the population of VSCT neurons increases in the course of the L-phase, reaching the maximum in the second half of the L-phase, and then rapidly decreases.

During the latent period of rhythmical scratching (when the pinna had already been stimulated, but rhythmical ENG (electroneurogram) bursts had not yet appeared), a tonic activation was observed in some neurons (Fig. 23A, see also Fig. 25 for decapitate preparation); these neurons are marked by filled circles in Fig. 24A (*stim* column). The resting discharge of some neurons was inhibited during the latent period (marked by minuses). Activity of the remaining neurons did not change (marked by open circles). One can see in Fig. 24A that the pattern of activity of individual VSCT neurons during the latent period is correlated with their phase distribution. The units, activated by pinna stimulation, prevail in the upper part of the graph, i.e., they fire in long bursts during the greater part of the L-phase. In contrast, the neurons which were inhibited or unaffected by stimulation of the pinna, prevail in the lower part of the graph, i.e., they fire in short bursts in the second half of the L-phase and in the S-phase. However, these groups are partly overlapping in the middle of the graph. These correlations will be discussed in the concluding section of this chapter.

The phase distributions of VSCT neurons projecting to the vermis and to the pars intermedia differ from each other, but the difference is not great (Arshavsky et al. 1978b).

The rhythmical activity of VSCT neurons can be also observed during fictitious scratching, in decapitate preparations, evoked by stimulation of the cervical spinal cord[4] (Fig. 25). The pattern of this activity is almost the same as in thalamic cats. In decapitate cats, the overall activity of VSCT neurons also increases during the L-phase and decreases

---

[4] Because of the spinal cord transection, in these experiments identification of neurons by their antidromic response to stimulation of the cerebellum was impossible. A less reliable criterium for identification was used, i.e., the antidromic response to stimulation of the contralateral lateral funiculus of the spinal cord where VSCT axons ascend (hatched area in Fig. 18C)

by the end of this phase and in the S-phase. The discharge frequencies of neurons in decapitate cats can reach 250–300 pulses $s^{-1}$ (Fig. 25) which is even more than in thalamic ones. Thus, the central spinal mechanism of scratching provides the intensive rhythmical activity of VSCT neurons.

Rhythmical activity of VSCT neurons arises not only when the scratch reflex is evoked on the ipsilateral (to a neuron) side of the spinal cord, but also during contralateral scratching. Figure 26 shows the activity of two VSCT neurons during ipsilateral (A, C) and contralateral (B, D) scratching. One can see that in the latter case the activity is weaker. Besides, the phase of activity during contralateral scratching can differ from that during the ipsilateral scratching as in the neuron in Fig. 26C, D.

Thus, two series of experiments (locomotion of animals with deafferented hindlimbs and fictitious scratch reflex) showed that VSCT neurons are rhythmically active in the absence of any rhythmical afferent signals from the limb receptors. It follows that VSCT neurons are rhythmically driven by central mechanisms. The crucial role in the control of VSCT neurons is played by the central spinal mechanism since the rhythmical activity in decapitate (spinal) preparations is not lower than in thalamic ones.

A question arises: from this: does the VSCT convey signals on the activity of the central spinal mechanism only, or also on the activity of the executive motor apparatus? To find the answer, the activities of VSCT neurons during actual and fictitious scratching were compared. Figure 27 shows the activites of two neurons, recorded first during actual scratching (A, C) and then during fictitious scratching (B, D). The discharge patterns did not change after the animals had been immobilized.

Thus, elimination of the rhythmical afferent inflow from limb receptors when the limb is deafferented (see Sec. 3b, this Chapter), or when the animal is immobilized, exerts no noticeable effects on the rhythmical activity of VSCT neurons. Therefore, even in actual movements, the VSCT neurons are driven mainly (or completely) by the central spinal mechanism. One can conclude that the VSCT monitors the activity of this mechanism.

## 4. Spino-Reticulo-Cerebellar Pathway

### a) General Characteristics

Reticulo-cerebellar fibres originate mainly from three nuclei: the lateral
and paramedian reticular nuclei of the medulla oblongata and the nucleus
reticularis tegmenti pontis (Jansen and Brodal 1954; Brodal 1957).
The spino-reticulo-cerebellar pathway (SRCP) relays mainly in the
lateral reticular nucleus (LRN) (Jansen and Brodal 1954; Oscarsson
1973). The fibres terminating in the LRN ascend mostly in the ipsilateral ventrolateral funiculus of the spinal cord (Brodal 1949, 1957;
Jansen and Brodal 1954; Morin et al. 1966; Künzle 1973) and form the
"bilateral ventral flexor reflex tract" (bVFRT) (Lundberg and Oscarsson
1962b; Grant et al. 1966; Oscarsson and Rósen 1966; Oscarsson 1967,
1973; Rósen, Scheid 1973a, b; Clendenin et al. 1974b).
    Axons of the LRN neurons terminate mostly in the ipsilateral
half of the cerebellar cortex as mossy fibres, a part of the reticulo-cerebellar fibres, especially those terminating in the vermis, from bilateral projections (Crichlow 1970; Clendenin et al. 1974a; Matsushita and Ikeda
1976). Reticulo-cerebellar fibres terminate not only in the cerebellar
cortex, but also give out collaterals terminating in the cerebellar nuclei
(see Chap. V).
    While studying degeneration of the mossy fibres in the anterior lobe of
the cerebellum after transection of various input pathways, Szenthagothai
(1964) came to the conclusion that reticulo-cerebellar fibres terminate
mainly in the superficial zone of the granular layer. This distinguishes
them from DSCT and VSCT mossy fibres which terminate mainly in the
deeper zone of the granular layer. On the basis of electrophysiological
studies Arshavsky et al. concluded that the fibres of direct spino-cerebellar tracts (DSCT and VSCT), on the one hand, and the reticulo-cerebellar
fibres, on the other, terminate on different types of granule cells (types
I and II, respectively) (Arshavsky et al. 1969a, b, 1970, 1971a, 1972c;
Arshavsky 1972). The type I granule cells exert excitatory action upon
all other cerebellar neurons including Purkinje cells (Fig. 13B). Therefore,
they directly participate in generating cerebellar output signals. The
type II granule cells (which are excited by reticulo-cerebellar fibres)
do not affect the Purkinje cells directly; by exciting the Golgi cells
(which inhibit granule cells), they influence the transmission of signals
coming through direct spino-cerebellar tracts. In Chap. IV and V we
shall consider the role of various inputs in generating the cerebellar
output signals.

Neurons of bVFRT (and, therefore, LRN neurons) are strongly affected by the flexor reflex afferents (Lundberg and Oscarsson 1962b; Grant et al. 1966; Oscarsson and Rosén 1966, Crichlow and Kennedy 1967; Oscarsson 1967, Bruckmoser et al. 1970b; Rosén and Scheid 1973a, b, c; Clendenin et al. 1974b). Their receptive fields are very wide. Most neurons respond to stimulation of the nerves of three to four limbs. Signals from different limbs converge, to a great extent, at the level of bVFRT neurons.

Besides the signals from the flexor reflex afferents, SRCP neurons receive signals from various supraspinal structures. The bVFRT neurons are monosynaptically excited by the vestibulo-spinal tract (Lundberg and Oscarsson 1962b; Grant et al. 1966, Oscarsson and Rosén 1966, Osarsson 1967, 1973; Grillner et al. 1968a; Rosén and Scheid 1973b; Clendenin et al. 1974b). Accordingly, it was demonstrated that spino-reticular neurons, sending axons to the LRN, respond to natural stimulation of vestibular receptors (Pompeiano 1975a, 1977). Furthermore, the transmission from the flexor reflex afferents to bVFRT neurons (as was also found for other pathways activated by the flexor reflex afferents) is facilitated by the pyramidal tract and inhibited by the reticulo-spinal tract. Thus, these descending pathways can affect the activity of SRCP neurons (Oscarsson 1973).

Still more numerous are the connections between the supraspinal centres and the LRN. As morphological and electrophysiological studies have shown, LRN neurons receive mono- and polysynaptic influences from vestibular nuclei (Ladpli and Brodal 1968), nuclei of the reticular formation (Brodal 1957), red nuclei (Walberg 1958; Courville 1968; Bruckmoser et al. 1970b; Kitai et al. 1974; Corvaja et al. 1977), fastigial and, to a lesser extent, interpositus nuclei of the cerebellum (Walberg and Pompeiano 1960; Bruckmoser et al. 1970b; Kitai et al. 1974; Corvaja et al. 1977), and from the cerebral cortex (Jansen and Brodal 1954; Brodal 1957; Brodal et al. 1967; Crichlow and Kennedy 1967; Bruck-moser et al. 1970a; Rosén and Scheid 1973a; Kitai et al. 1974). It follows that organization of the SRCP afferent connections is very similar to that of the VSCT afferents. Like the VSCT neurons, the spino-reticular neurons receive signals mostly from the flexor reflex afferents, their receptive fields are very large, and they are strongly influenced by supraspinal centres. According to the hypothesis proposed by Lundberg and Oscarsson, spino-cerebellar neurons with such pro-perties convey information not on the activity of the executive motor apparatus, but on the activity of the spinal motor centres.

Besides the bVFRT, there is spino-reticulo pathway terminating in the LRN and transmitting signals related to the ipsilateral forelimb. This input to the LRN was first investigated in Oscarsson's laboratory

(Clending et al. 1974c), and a hypothesis was advanced that it monitors the activity of spinal motor centres. Signals concerning the forelimb activity are transmitted to the LRN, at least partly, by the population of propriospinal neurons from the C3–C4 segments which were extensively studied in Lundberg's laboratory (Illert et al. 1978; Alstermark et al. 1981). These neurons have bifurcating axons, the descending collaterals of the axons projecting to forelimb motoneurons and the ascending ones to the LRN (Fig. 29 A–C). The neurons receive inputs from the forelimb and neck afferents as well as from supraspinal structures (cerebral cortex, red nuclei, bulbar reticular formation).

### b) Activity of LRN Neurons During Locomotion and Scratching

The LRN neurons were identified by the antidromic response to stimulation of the hindlimb zone in the cerebellar anterior lobe (Fig. 28 A–D). During locomotion LRN neurons fire in rhythmical bursts

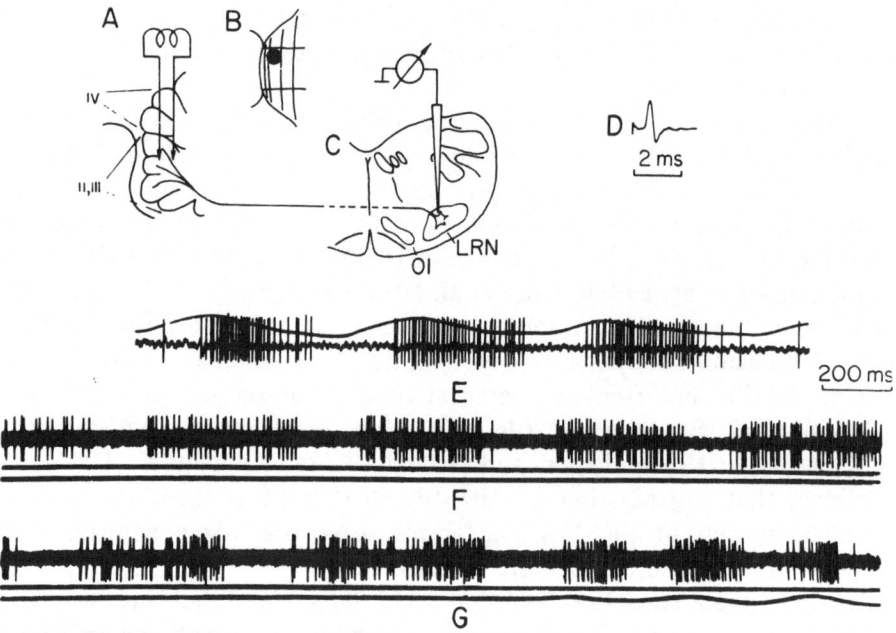

**Fig. 28 A–C.** Recording and identification of LRN neurons. Neurons were recorded from the lateral reticular nucleus and identified by the antidromic response (**D**) to stimulation of the hindlimb area in the vermis of the cerebellar anterior lobe. **E–G** Activity of two LRN neurons (**E** and **F, G** respectively) during locomotion. In **E** activity of the neurons was recorded together with movements of the ipsilateral forelimb, in **F, G** with movements of the forelimb (*middle trace*) and hindlimb (*lower trace*). Protraction of the limb corresponds to the upward deflection of the beam (Arshavsky et al. 1974)

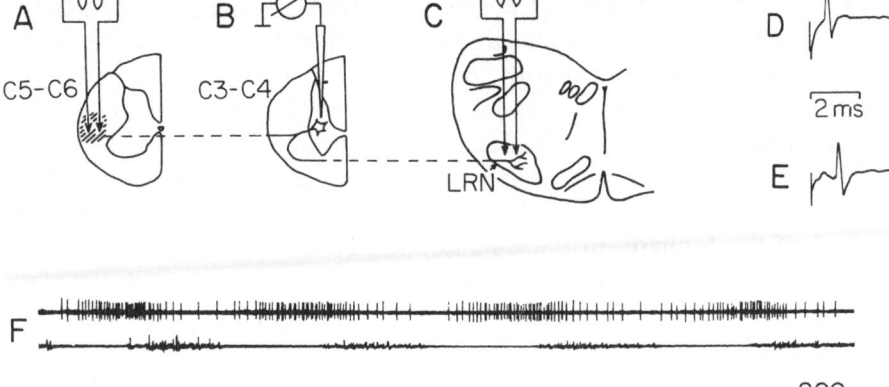

Fig. 29 A–E. Recording and identification of spino-reticular neurons. Neurons were recorded from the *C3–C4* segments (**B**) and identified by the antidromic response to stimulation of their descending axons in the *C5–C6* segments (**A, D**) and to stimulation of their axonal terminations in the LRN (**C, E**). **F** Activity of a spino-reticular neuron during fictitious locomotion of forelimbs. The *lower trace* is the ENG of the ipsilateral n.triceps. **G** Activity of a LRN neuron during fictitious locomotion of forelimbs. The *lower trace* is the ENG of the ipsilateral n.biceps (Arshavsky et al. 1985)

related with stepping movements, the discharge frequencies being 100–200 (sometimes up to 300–400) pulses s$^{-1}$ (Fig. 28E) (Arshavsky et al. 1974). As in VSCT neurons, in some LRN neurons rhythmical activity appeared before any stepping movements began (Fig. 28F, G). This finding suggested that rhythmical activity was determined by central mechanisms of locomotion which became active prior to the appearance of limb movements. This was proven while studying the activity of LRN neurons during fictitious locomotion. In these experiments, fictitious locomotion of the forelimbs was evoked by stimulation of the mesencephalic locomotor region. To facilitate the locomotor activity, the spinal cord was transected at the lower thoracal level (Yamaguchi 1982). Figure 29G shows how during fictitious locomotion activity of a LRN neuron is rhythmically modulated in relation to efferent activity. The same is true for spino-reticular neurons belonging to the population of propriospinal neurons from the C3–C4 segments studied in Lundberg's laboratory. Figure 29F shows rhythmical discharges of such a neuron, giving an axon to the LRN, during forelimb fictitious locomotion.

Evidence for the central drive to LRN neurons from the spinal hindlimb centre was obtained while studying their activity during ficti-

tious scratching (Arshavsky et al. 1977, 1978a). During ipsilateral fictitious scratching, LRN neurons fire in rhythmical bursts which, in most
units, more or less coincide with the bursts in the extensor nerve (Figs.
30; 32A, C, E; 33; 34A, C). As in locomotion, the discharge frequency
within the bursts can reach 100-200 (sometimes up to 400) pulses $s^{-1}$.

Figure 31A shows the phase distribution of LRN neurons in the
cycle of fictitious scratching. One can see that most units are active at
the end of the L-phase, in the S-phase, and at the very beginning of the
subsequent L-phase. Only a few neurons are active during the greater
part of the L-phase. Figure 31B shows the frequency curve for the
"average" LRN neuron. One can see that the overall activity of the
population of LRN neurons increases by the end of the L-phase and
sharply decreases after the end of the S-phase.

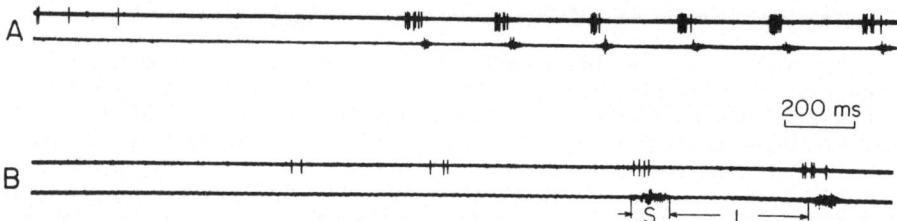

Fig. 30A, B. Activity of two LRN neurons during fictitious scratching. The *lower trace*
is the gastrocnemius ENG (Arshavsky et al. 1978a)

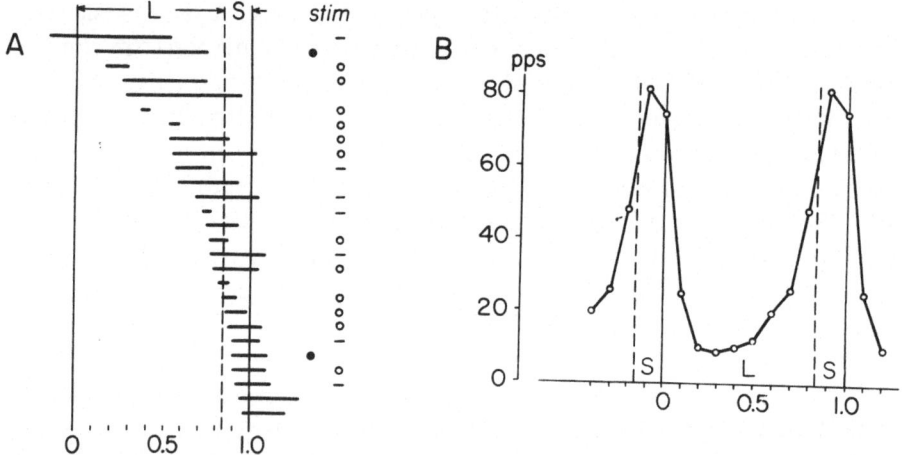

Fig. 31A, B. Relationship between the activity of LRN neurons and the phase of
the scratch cycle. A The phase distribution of 27 LRN neurons. In the column
*stim*, the behaviour of most neurons during the latent period of scratching is shown
(*filled circles*, excitation; *open circles*, no change; *minus*, inhibition of the activity).
B The discharge frequency of the "average" LRN neuron as a function of the
phase of the cycle (Arshavsky et al. 1978a)

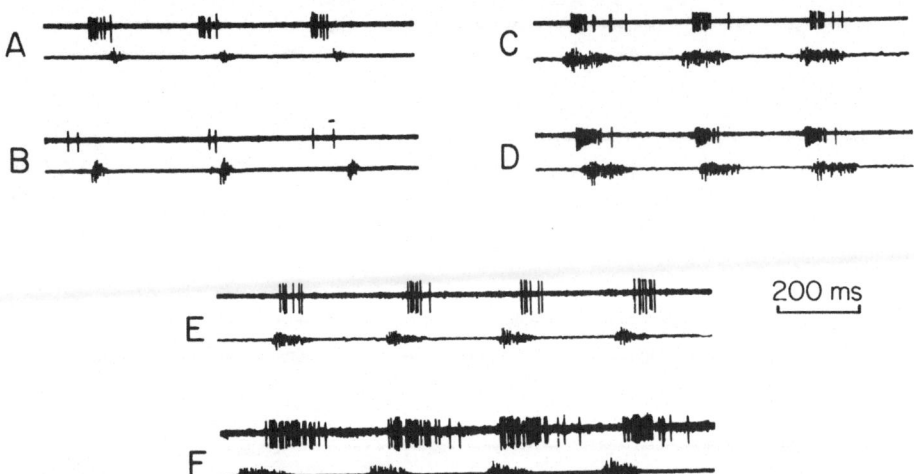

**Fig. 32 A—F.** Activity of three LRN neurons during ipsilateral (**A, C** and **E**) and contralateral (**B, D** and **F**) fictitious scratching. The *lower trace* is the ENG of ipsilateral (**A, C** and **E**) and contralateral (**B, D** and **F**) n.gastrocnemius (Arshavsky et al. 1978a)

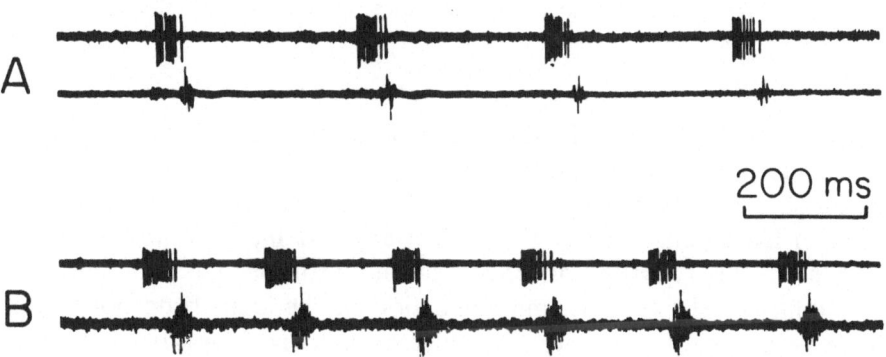

**Fig. 33A, B.** Activity of a LRN neuron recorded initially during actual scratching (**A**) and then during fictitious scratching (**B**) after Flaxedil injection (5 mg kg[-1], i.v.). In **A**, the *lower trace* is the gastrocnemius EMG, and in **B**, the ENG (Arshavsky et al. 1978a)

During the latent period of scratching, activity of the majority of the LRN neurons either does not change (such neurons are marked by open circles in the *stim* column in Fig. 31A) or is inhibited, like that of the cell in Fig. 30A (such neurons are marked by minuses).

When fictitious scratching was evoked on the opposite side, LRN neurons also fired in rhythmical bursts (Fig. 32B, D, F). On average, the rhythmical activities during ipsi- and contralateral scratching were the same. Therefore, while either of the two spinal hindlimb centres is

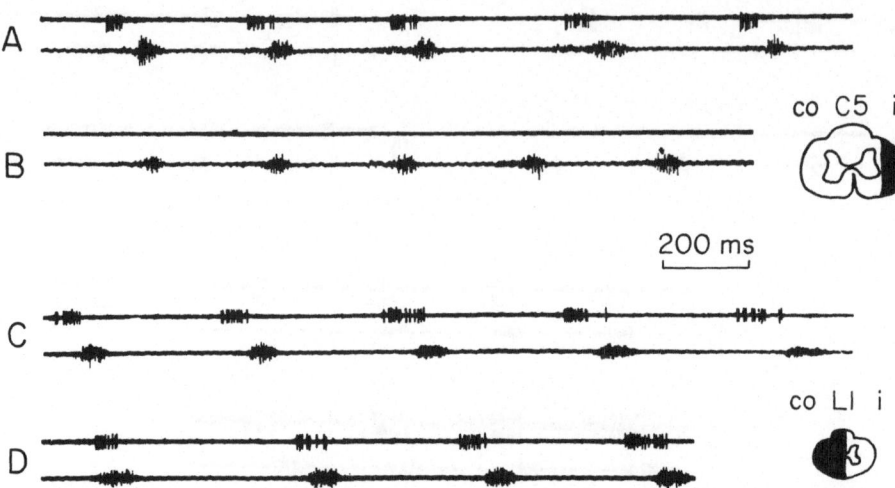

**Fig. 34 A–D.** Effects of partial transections of the spinal cord on the activity of LRN neurons during fictitious scratching. **A, B** Activity of a LRN neuron recorded before (**A**) and after (**B**) transection of the ipsilateral lateral funiculus at the *C5* level. **C, D** Activity of a LRN neuron recorded before (**C**) and after (**D**) contralateral hemisection of the spinal cord at the L1 level. The *lower trace* is the grastrocnemius ENG. The extent of the lesion is shown as a *black area* on the spinal cord cross-section. *Co* and *i* – contralateral and ipsilateral (in relation to a neuron) sides of the spinal cord (Arshavsky et al. 1978a)

rhythmically active, both spino-reticulo-cerebellar pathways convey similar signals to the cerebellum.

These data clearly show that LRN neurons can be rhythmically active in the absence of any rhythmical afferent input. Moreover, comparison of the activities of LRN neurons during actual and fictitious scratching has shown that immobilization of the animals do not affect the pattern of neuron activity. The unit shown in Fig. 33 was first recorded during actual and then during fictitious scratching. It can be noted that its activity practically does not change after immobilization of the animal. This means that even in actual scratching afferent signals from the moving limbs have almost not influence upon SRCP neurons.

According to the morphological and electrophysiological data described in the previous section, the LRN neurons receive signals both from the spinal cord and from various supraspinal centres. In subsequent chapters it will be shown that during locomotion and scratching, cerebellar and brain stem neurons are rhythmically active and thus capable of exerting rhythmical influence upon LRN neurons. To determine the contribution of supraspinal inputs to the rhythmical modulation of LRN neurons, the following experiments were carried out. The ipsilateral lateral funiculus of the spinal cord (in which spino-reticular fibres ascend) was transected[5] and effects of the transection upon activity of LRN neurons during fictitious scratching were studied. As will be shown later (Chap. IV and V), after such a transetion the rhythmical activities of brain stem and cerebellar neurons persist. However, the transection leads to total cessation of the rhythmical activity of LRN neurons. This can be seen in Fig. 34A, B which shows the activity of a LRN neuron before (A) and after transection (B).

It follows that rhythmical activity of LRN neurons is determined by the signals coming via the spino-reticular fibres. However, the neurons giving rise to the bVFRT are themselves also affected by different supraspinal structures (see preceding section). The question remains whether the rhythmical activity of the spino-reticular neurons is determined by supraspinal influences. To answer this question, a contralateral hemisection of the spinal cord was performed. Such an operation almost completely eliminates any rhythmical supraspinal influences upon the bVFRT neurons since: (1) all the pathways descending in the contralateral half of the spinal cord are transected; (2) rhythmical modulation in ipsilateral descending tracts becomes much weaker as a result of the VSCT transection (see Chap. IV). Figure 34C, D shows the activity of a LRN neuron recorded during fictitious scratching before (C) and after (D) contralateral hemisection. One can see that the firing pattern does not change. This means that rhythmical activity of the bVFRT neurons does not depend on any rhythmical supraspinal influences either.

While studying behaviour of the LRN neurons during locomotion and scratching, it was shown that they are rhythmically active both in animals with an intact cerebellum and in the decerebellate ones. Since removal of the cerebellum causes cessation of rhythmical activity in the brain stem neurons (see Chap. III), this finding also means that supraspinal centres do not play any significant role in the rhythmical control of LRN neurons.

Thus, the body of facts presented in this section points out that the SRCP, like VSCT, conveys information concerning activity of the central spinal mechanisms. Both the afferent signals from the receptors of the executive motor apparatus and the signals from the supraspinal centres are of minor importance for the rhythmical activity of the SRCP neurons during locomotion and scratching.

---

[5] Transection of the ipsilateral lateral funiculus was performed at the C4–C5 level. This operation does not eliminate scratching since in the cervical spinal cord the pathway of the scratch reflex descends mainly on the contralateral side (Deliagina 1977)

## 5. VSCT and SRCP Convey Messages About Activity in the "Leading" Region of Lumbo-Sacral Spinal Cord

The data presented in the preceding sections have demonstrated that both the VSCT and SRCP convey messages on the activity of intraspinal mechanisms controlling stepping and scratching movements. Earlier (see Chap. I, Sec. 6) it was shown that the lumbo-sacral part of the spinal cord, which contains the spinal hindlimb centres, is not functionally homogeneous. During scratching, the cycle duration as well as the durations of the flexor and extensor phases are determined by the neuronal mechanisms of the L3–L5 segments. This region is the '"lead-ing" one since it establishes the main parameters of rhythmical activity in the whole lumbo-sacral spinal cord. Some experimental data suggest that approximately the same region determines the main parameters of stepping movements of the hindlimbs as well, because this region is necessary for generating the locomotor rhythm.

The role of the "leading" region of the lumbo-sacral spinal cord in generating rhythmical signals conveyed by the VSCT and SRCP was studied during fictitious scratching (Arshavsky et al. 1984a). In this study, the method of switching off the caudal spinal segments by means of local cooling (described in Chap. 1, Sect. 6) was used. The L5 segment was cooled, and the activity of VSCT neurons from the L4 segment as well as of LRN neurons was recorded. Figure 35 shows activities of the VSCT neuron (A, B) and of the LRN neuron (C, D) during fictitious scratching before (A, C) and then during cooling (B, D). Cooling results in cessation of the rhythmical activity in the L5 and more caudal segments of the spinal cord which is proven by the absence of any rhythmical bursts in n.gastrocnemius motoneurons (which are located in the L7 and S1 segments) and by reduction of the bursts in n.sartorius (whose motoneurons are located in the L4 and L5 segments). One can see from Fig. 35 that cooling does not affect the rhythmical activity of the VSCT and LRN neurons in spite of the fact that, during cooling, the overwhelming majority of the interneurons and moto-neurons of the spinal hindlimb centre do not participate in the rhythmical activity. Similar results were obtained while recording VSCT neurons from the L5 segment and cooling the L6 segment: rhythmical activity of most VSCT neurons did not change during cooling (it should be called to mind that most VSCT neurons are located in the L4 and L5 segments).

This finding means that both the VSCT and the SRCP convey messages on the activity of neuronal mechanisms of the "leading" region of the hindlimb centre, i.e., on the activity of the spinal generator of rhythmical oscillations. In contrast, the processes taking place in the

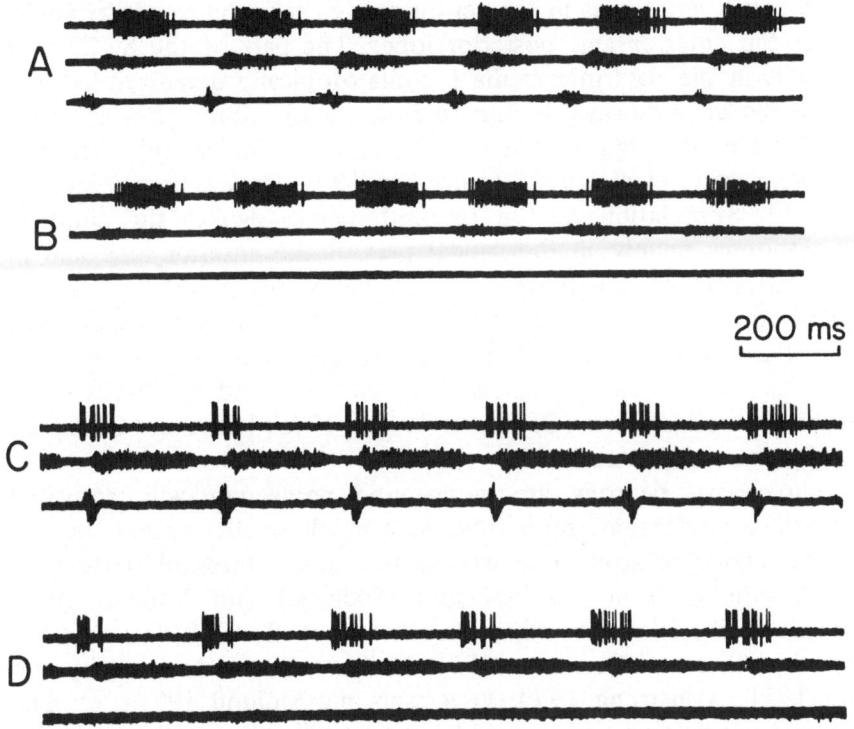

**Fig. 35 A—D.** Effects of cooling of the L5 segment upon activity of a VSCT neuron (**A, B**) and a LRN neuron (**C, D**) during fictitious scratching. Activity of the neurons is shown before cooling (**A, C**) and 60 s after beginning of cooling (**B, D**) when the caudal part of the spinal cord is "switched off" as can be determined from cessation of the rhythmical activity in n.gastrocnemius (*lower trace*). The *middle trace* is the sartorius ENG. For methods see Fig. 9 (Arshavsky et al. 1984a)

caudal part of the lumbo-sacral enlargement containing the overwhelming majority of motoneurons and interneurons of the hindlimb centre (spinal "output" mechanisms) are not monitored by the VSCT and SRCP.

## 6 Spino-Olivo-Cerebellar Pathway

The SOCP terminates in the cerebellar cortex as climbing fibres (see Sect. 1). Signals from the spinal cord reach the inferior olives via several groups of fibres ascending in different funiculi of the cord (Oscarsson 1967, 1969a, b, 1973; Larson et al. 1969a; Miller and Oscarsson 1970; Oscarsson and Sjölund 1977a, b, c). Before entering the cerebellum, the olivo-cerebellar fibres cross the midline.

The SOCP terminates in the vermis as well as in the pars intermedia, both in the anterior and posterior lobes. The part of the SOCP that terminates in the pars intermedia is somatotopically organized like the spino-cerebellar pathways terminating as mossy fibres (Eccles et al. 1968; Kitai et al. 1969; Oscarsson 1969a, 1973; Miller and Oscarsson 1970; Armstrong et al. 1971; Armstrong 1974; Oscarsson and Sjölund 1977a, b). Stimulation of the forelimb nerves evokes the climbing fibre responses mainly in the caudal part of the anterior lobe and in the rostral part of the paramedian lobule, while stimulation of the hindlimb nerves evokes responses in the rostral part of the anterior lobe and in the caudal part of the paramedian lobule (hatched areas in Fig. 13A). In the vermis, signals from the fore- and hindlimbs reach several longitudinal zones (Oscarsson 1969a, 1973; Miller and Oscarsson 1970; Oscarsson, Sjölund 1977a, b, c).

Spino-olivary neurons are mono- und polysynaptically activated by the flexor reflex afferents and, to a much smaller extent, by the muscle afferents of group I as well as by the low threshold cutaneous afferents coming from the foot pads (Sedgwick and Williams 1967; Oscarsson 1967, 1969a, b, 1973; Armstrong et al. 1968; Larson et al. 1969a, b; Miller and Oscarsson 1970; Eccles et al. 1971a; Murphy and Sabah 1971; Armstrong 1974; Oscarsson and Sjölund 1977a, c). The excitatory convergence is very extensive: most units can be activated from all the nerves in one or more limbs. Studies of the reactions of Purkinje cells to natural stimulation of cutaneous receptors have also shown that the olivar neurons have very large receptive fields (Leicht et al. 1977). The wide convergence of the signals from different nerves is, to a great extent, realized already at the level of the spino-olivary neurons.

Polysynaptic transmission of signals from the flexor reflex afferents to spino-olivary neurons is influenced by supraspinal mechanisms (Sjölund 1978). The transmission is inhibited by the reticulo-spinal system and facilitated by the cortico-spinal and rubro-spinal tracts. On the other hand, supraspinal centres do not influence the spino-olivary neurons which are monosynaptically linked to the flexor reflex afferents. The neurons of the inferior olives (in which the SOCP is relayed) are strongly affected by various brain centres: the cerebral cortex, the caudate nucleus, and some mesencephalic structures (Crill and Kennedy 1967; Sedgwick and Williams 1967; Miller et al. 1969a, b; Oscarsson 1973; Armstrong 1974).

Signals coming via the climbing fibres are often considered as being of great importance for the functioning of the cerebellum (cf. Marr 1969; Eccles 1973, 1977; Armstrong 1974, 1978; Llinás 1974; Boylls 1975). The reason for this viewpoint is evident. Each Purkinje cell is innervated by a single climbing fibre exerting a strong excitatory influence on it (see Sect. 1). Therefore, the system of climbing fibres can efficiently and selectively control the activity of Purkinje cells. In particular, it was suggested that the signals coming through the SOCP are especially important for the control of locomotion (Armstrong 1974; Boylls 1975; Rushmer et al. 1976). However, the recording of inferior olive neurons during locomotion in thalamic cats demonstrated that the activity of the olivo-cerebellar neurons was not correlated (or only slightly correlated) with stepping movements (Boylls 1977). Corresponding results were obtained while studying the activity of Purkinje cells during locomotion: "complex spikes" (reflecting the climbing fibre input) usually were not related to movements (Orlovsky 1972d, see Chap. V). However, later Udo and colleagues (Udo et al. 1981) found a correlation between the appearance of "complex spikes" and the phase of stepping movements, but the correlation was not so pronounced as that for "simple spikes" generated by the mossy fibre input.

Corresponding results were obtained while recording the activity of Purkinje cells in monkeys trained to perform simple rhythmical movements with the arm (Thach 1968; Mano 1974). It was demonstrated that the "complex spikes" occurred irregularly and in no obvious relation to the movements. On the other hand, in a more complicated situation (quick arm movement in response to a light signal), the complex spikes appeared in relation to movements. However, complex spikes frequently occurred not after the beginning of movement but before or simultaneously with it, which suggests that they were determined by the signals coming via the cerebro-olivo-cerebellar pathway rather than via the SOCP.

Since one of the main inputs to the olivary neurons is that from the flexor reflex afferents, Oscarsson and colleagues suggested that the SOCP (like the VSCT and SRCP) monitors the activity of spinal motor centres (Miller and Oscarsson 1970; Oscarsson 1973; Oscarsson and Sjölund 1977c; Andersson and Sjölund 1978; Sjölund 1978; Ekerot et al. 1979). To test this hypothesis, the activity of olivo-cerebellar neurons was recorded during fictitious scratching. The scheme of the recording and the identification of neurons was similar to that shown in Fig. 28 A–C. All one could observe during scratching was a minor increase of the discharge frequency; any rhythmical activity, related to the efferent activity, was absent (Fig. 36).

Thus, the problem concerning the nature of signals conveyed by the SOCP remains unsolved. In the movements studied so far, the SOCP does not participate in transmitting the signals either on the

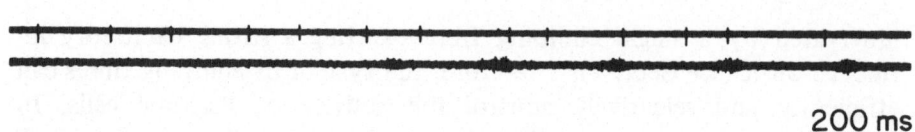

200 ms

Fig. 36. Activity of an inferior olive neuron during fictitious scratching. The *lower trace* is the gastrocnemius ENG (Arshavsky and Pavlova, unpublished)

activity of the executive motor apparatus or on the activity of the central spinal mechanisms. It was suggested by Marr (1969) that signals conveyed by the SOCP do not participate in the current regulation of movements; they are used only during the organization of motor programs. Some experimental data supporting this point of view were obtained in Ito's laboratory (Ito 1982; Ito and Kano 1982).


## 7  Conclusion

Studies of the activity of neurons, belonging to the spino-cerebellar pathways, during locomotion and scratching have shown that they convey two types of signals. The DSCT transmits information on the activity of the executive motor apparatus, above all, on the phase and intensity of contraction of single muscles. Cessation of the rhythmical afferent signals from the receptors of the executive motor apparatus (transection of dorsal roots in locomotion or immobilization of the animal in scratching) eliminates the rhythmical activity of DSCT neurons. In contrast, the VSCT and SRCP convey information concerning the activity of the central spinal mechanisms. Cessation of the rhythmical afferent signals from the limb receptors has no influence upon the rhythmical activity of the neurons of these two pathways. This is clearly demonstrated by comparion of the activity of VSCT and LRN neurons during actual and fictitious scratching (see Figs. 27 and 33). It follows that the intraspinal processes must be considered as the main source of rhythmical activity of VSCT and SRCP neurons, irrespective of whether the spinal cord receives any afferent signals from the receptors of the moving limbs or not.

The conclusion that intraspinal processes are the main (if not the only) source of the rhythmical activity of VSCT and SRCP neurons requires some more consideration. As has mentioned, at rest both the VSCT and SRCP neurons easily respond to various peripheral stimuli, for example, to passive limb movements. Therefore, one could expect afferent inflow to the spinal cord during active limb movements also affect the neurons. Since this is not the case,

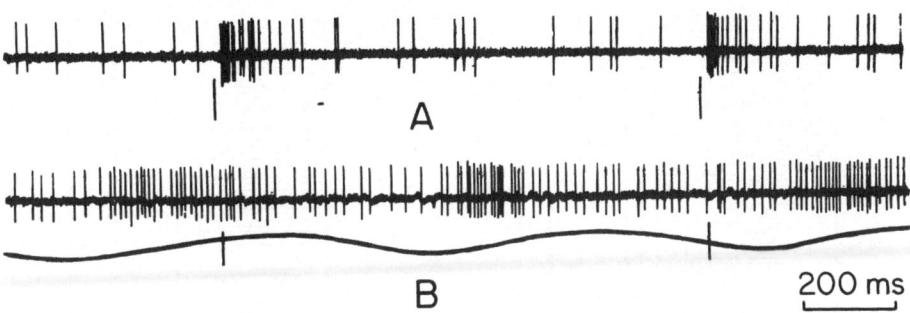

**Fig. 37A, B.** Responses of a LRN neuron to stimulation of the ipsilateral common radial nerve at rest **(A)** and during locomotion **(B)**. Moments of stimulation are marked by *vertical lines*. The *lower trace* in **B** shows movements of the ipsilateral forelimb (protraction — up) (Arshavsky et al. 1974)

afferent influences upon the VSCT and SRCP neurons during active movements must be suppressed by a certain mechanism. Such suppression of the peripheral inputs was directly demonstrated while studying the activity of LRN neurons during locomotion (Arshavsky et al. 1974). Figure 37 shows that responses of a LRN neuron to stimulation of the peripheral nerve are strongly present at rest (A) while they are hardly visible during locomotion (B). This result can be explained in the following way. The SRCP neurons, like the VSCT ones, receive a strong excitatory input from the flexor reflex afferents. Meanwhile, the flexor reflex is subject to dramatic changes when the spinal locomotor mechanism comes into operation. It was demonstrated in Lundberg's laboratory that the injection of DOPA, i.e., the drug activating the spinal locomotor mechanism, results in inhibition of interneurons mediating the normal (short latency) flexion reflex (Anden et al. 1966; Lundberg 1981). If we suppose that these interneurons also mediate the influences of the flexor reflex afferents upon the SRCP neurons, the decrease of efficiency of peripheral inputs to LRN neurons during locomotion can be easily explained.

In the experiments with cooling of the spinal cord, it was found that during scratching, the VSCT and SRCP convey messages on the activity of the spinal generator of rhythmical oscillations which determines the temporal pattern of scratching (the rhythm of activity and duration of the flexor and extensor phases) for the whole spinal hindlimb centre. What kind of information on the activity of the generator is conveyed by the VSCT and SRCP? To answer this question, we shall compare the activity of VSCT and SRCP neurons during scratching with that of spinal interneurons from the L4 and L5 segments described in Sect. 6, Chap. I. Figure 38A shows the phase distribution

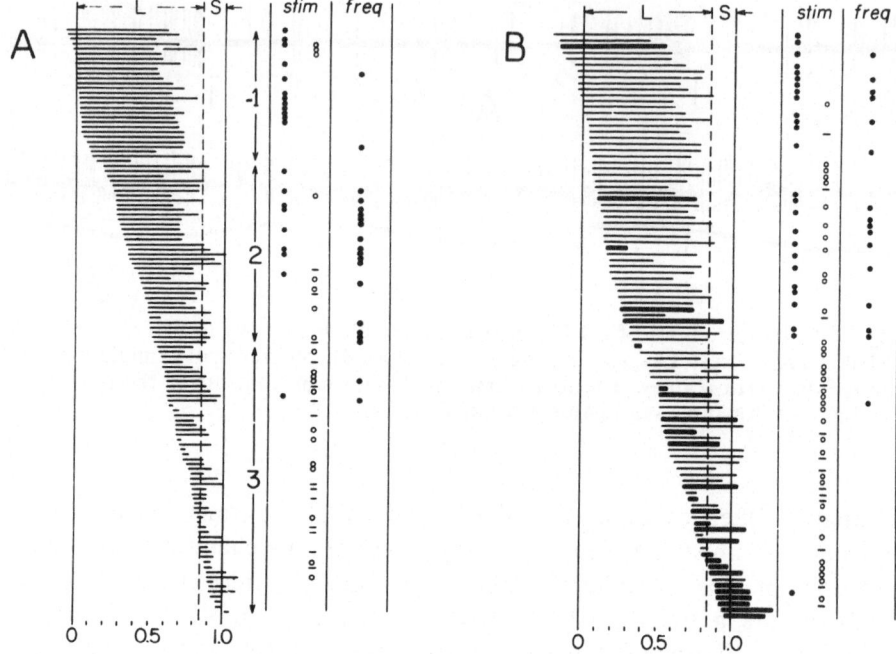

**Fig. 38A, B.** Comparison of behaviour of spinal interneurons and neurons of spino-cerebellar pathways during fictitious scratching. **A** The phase distribution of spinal interneurons from L4 and L5 segments (repetition of Fig. 11). **B** The phase distribution of VSCT neurons (*thin lines*) and LRN neurons (*heavy lines*) (combination of Figs. 24A and 31A). Abbreviations as in Fig. 11 (Arshavsky et al. 1978b)

of spinal interneurons (repetition of Fig. 11), and Fig. 38B – the phase distribution of both VSCT neurons (thin lines) and LRN neurons (heavy lines) (the graph is a combination of Figs. 24A and 31A). For most units, their behaviour during the latent period of rhythmical scratching is also indicated (the *stim* column); the units with a discharge rate increasing in the course of the burst are marked in the column *freq*. There is a striking similarity between the behaviour of the spinal interneurons and that of the neurons of the spino-cerebellar pathways. In both cases, the neurons are continuously recruited throughout the cycle, while the termination of activity in most neurons occurs near the end of the cycle. The neurons which are tonically facilitated during the latent period of scratching prevail in the upper parts of graphs A and B, those which are inhibited or unaffected, in the lower parts. Finally, the discharge frequency of a large number of neurons situated in the upper and, especially, in the middle parts of the graphs increases in the course of the burst.

We think that this correlation is not accidental; it probably means that interneurons from the "leading" region of the spinal hindlimb centre generating the temporal pattern of scratching simultaneously drive the neurons of spino-cerebellar pathways. As a result, the latter convey to the cerebellum information on the activity of neuronal mechanisms of the spinal rhythmical generator.

In Chap. I, while considering the phase distribution of spinal interneurons during scratching, these neurons were divided into three groups according to the phases of their activity in the cycle, and the possible roles of interneurons of different groups in the process of rhythmical generation were discussed. These groups are indicated in Fig. 38A. Each of the groups has its "image" in the population of VSCT and LRN neurons (Fig. 38B). One can suppose that the spino-cerebellar neurons situated in the upper part of the graph (almost exclusively the VSCT ones) convey messages on activity of the group 1 interneurons. The neurons from the middle part of the graph (mainly the VSCT ones) convey messages on activity of the group 2 interneurons. Finally, the neurons situated in the lower part of the graph (both the VSCT and LRN ones) convey messages on the activity of the group 3 interneurons.

The data described in this chapter fully confirm the remarkable idea of Lundberg and Oscarsson that ascending pathways can convey information on intraspinal processes. On the other hand, the same data evoke serious doubts concerning the concrete interpretation of this idea given by Lundberg (1971) who supposed that the VSCT conveys information on transmission in inhibitory pathways impinging upon motoneurons. Indeed, in our studies it was demonstrated that the VSCT and SRCP convey messages only on the activity of neuronal mechanisms of the L3—L5 segments; therefore, activity of the majority of interneurons of the spinal hindlimb centre (including the inhibitory ones) is not reflected in the activity of spino-cerebellar neurons. Besides, it follows from the comparison of Figs. 38A and B that the activity of VSCT and LRN neurons reflects activity of the whole population of rhythmically modulated interneurons from the L3—L5 segments, and not only of a single group. Thus, it seems more likely that the VSCT and SRCP convey to the cerebellum information on the activity of all the main groups of interneurons from the "leading" region of the spinal hindlimb centre.

# III  Signals Conveyed by Descending Tracts

In Chap. II the rhythmical signals conveyed to the cerebellum by ascending pathways during locomotion and scratching were described. Here, the role of these input signals for generating the cerebellar output signals will be considered.

Fibres originating from the cerebellum terminate in various structures of the brain stem, in particular, in the vestibular, reticular and red nuclei, which give rise to the vestibulo-spinal (VS), reticulo-spinal (RS) and rubro-spinal (RbS) tracts. Therefore, signals conveyed by descending tracts reflect cerebellar output signals (Fig. 1). In this chapter we shall describe the activity of VS, RS and RbS neurons during locomotion and scratching as well as the effects produced by cerebellar ablation upon this activity. Furthermore, we shall consider the effects produced in the spinal cord by signals coming through descending tracts.

## 1.  Vestibulo-Spinal-Tract

### a)  General Characteristics

Morphology and functional properties of the VS tract have been extensively studied (Brodal et al. 1962; Brodal 1969, 1974; Grillner and Hongo 1972; Wilson 1972; Kostyuk 1973; Fanardjian 1975; Pompeiano 1975b; Shapovalov 1975a, b; Wilson and Peterson 1978; Fukushima et al. 1979). At the end of the previous century it was found (Monakow 1883) that the VS tract originates mainly from the lateral vestibular nucleus of Deiters[6]. The nucleus contains cells of different size, including giant multipolar cells. Both large and small

---

[6] The VS tract orginating from Deiters' nucleus is sometimes called "the lateral VS tract" (Nyberg-Hansen 1964, 1966). A considerably smaller amount of VS fibres originate from the medial and descending vestibular nuclei. These fibres form the medial and recently described (Peterson et al. 1978a) caudal VS tracts, which remain outside the scope of the present book

neurons give rise to VS fibres. The VS tract descends ipsilaterally along the ventral part of the spinal cord down to the lumbo-sacral segments (Nyberg-Hansen and Mascitti 1964; Nyberg-Hansen 1966). The VS fibres terminate in the ventral horn of the spinal cord in Rexed's (1954) laminae VIII and VII; some fibres reach lamina IX and terminate on dendrites of motoneurons. The conduction velocity of VS axons ranges from 24 to 140 m s$^{-1}$, with a mode at 80–100 m s$^{-1}$ (Ito et al. 1964a; Wilson et al. 1967; Wilson and Yoshida 1969; Fanardjian, Sarkissian 1980).

The VS tract is somatotopically organized. The cells sending axons to the lumbo-sacral segments (i.e., to the spinal hindlimb centre) are located mainly in the dorso-caudal part of Deiters' nucleus, while the cells sending axons to the cervical segments (i.e., to the forelimb centre) – in its ventro-rostral part (Pompeiano and Brodal 1957a; Ito et al. 1964a; Nyberg-Hansen and Mascitti 1964; Wilson et al. 1967; Fanardjian and Sarkissian 1980). However, zones of the fore- and hindlimbs are partly overlapping; some of the VS fibres terminate in both the cervical and lumbo-sacral segments (Abzug et al. 1974).

Most of the VS neurons discharge spontaneously with frequencies 5–50 pulses s$^{-1}$ (Pompeiano and Cotti 1959; Orlovsky 1972b; Arshavsky et al. 1978c); the mean value in decerebrate cats being 15 pulses s$^{-1}$. Influences of the VS tract upon the spinal cord were studied by methods of destruction and stimulation of Deiters' nucleus or the VS tract. The results indicate that the VS tract exerts an excitatory influence mainly on extensor motoneurons. Lesion of Deiters' nucleus or transection of the VS tract results in a reduction of extensor tonus in the ipsilateral limbs (Sherrington 1898; Bach and Magoun 1947; Lindsley 1952; Gernandt and Thulin 1953; Sprague and Chambers 1953; Batini et al. 1957) as well as in a decrease of rhythmical extensor activity during locomotion (Orlovsky 1972a; Yu and Eidelberg 1981). Stimulation of Deiters' nucleus results in an increase of the activity and reflex excitability of extensor motoneurons of the ipsilateral limbs (Sprague et al. 1948; Thulin 1953; Pompeiano 1960; Brodal et al. 1962; Sasaki et al. 1962). This stimulation gives rise to monosynaptic EPSP's mainly in the knee and ankle extensor alpha- and gamma-motoneurons, and to a smaller extent in the hip and toe motoneurons, while polysynaptic EPSP's can be observed in almost all motoneurons of the hindlimb extensors (Lund and Pompeiano 1965, 1968; Grillner et al. 1966a, b; 1968b, 1969, 1970; Grillner 1969; Wilson and Yoshida 1969; Grillner and Hongo 1972; Kostyuk 1973; Shapovalov 1975a, b). Besides, stimulation of Deiters' nucleus results in facilitation of interneurons mediating the crossed extensor reflex (Bruggencate et al. 1969) and of interneurons in the Ia inhibitory pathway to flexors (Grillner et al. 1966a).

Stimulation of Deiters' nucleus also evokes polysynaptic EPSP's mainly in the extensor motoneurons on the contralateral side of the spinal cord (Grillner and Hongo 1972; Hongo et al. 1975).

Deiters' nucleus receives afferent signals from a number of sources. In particular, the signals from the spinal cord reach the nucleus through collaterals of the fibres of spino-cerebellar pathways; Deiters' nucleus being supplied by collaterals of both the mossy fibres (mostly of the DSCT) and the climbing ones (Lorente de No 1933; Eccles et al. 1967; Ito et al. 1969; Matsushita and Ikeda 1970a; b; Allen et al. 1972a, b; Andersson and Oscarsson 1978a). Besides, special spino-vestibular fibres seem to terminate in Deiters' nucleus (Brodal et al. 1962; Wilson et al. 1966; Wilson 1972).

However, the main flow of signals from the spinal cord reaches Deiters' nucleus through the cerebellum which is the main source of afferent fibres terminating on the VS neurons. Axons of both Purkinje cells and of neurons of the fastigial nucleus terminate in Deiters' nucleus (Allen 1924; Jansen and Brodal 1954; Walberg and Jansen 1961; Brodal et al. 1962; Eccles et al. 1967; Voogd 1964). The cerebello-vestibular pathway formed by the axons of Purkinje cells arises from the vermis (mostly from its lateral part) and terminates in the ipsilateral Deiters nucleus, mostly on the larger VS neurons. This pathway is topographically organized: the ventro-rostral part of the nucleus, giving projections to the cervical segments of the cord, receives fibres mainly from the forelimb area of the vermis, the dorso-caudal part, giving projections to the lumbo-sacral segments, mainly from the hindlimb area (Brodal et al. 1962; Pompeiano 1967; Fanardjian and Sarkissian 1980).

Since Purkinje cells are inhibitory neurons (see. Chap. II), stimulation of the vermis results in monosynaptic IPSP's in the VS neurons (Ito and Yoshida 1964, 1966; Ito 1965, 1967; Ito et al. 1966, 1968a; Akaike et al. 1973). Stimulation of the spinal cord or peripheral nerves results in complex synaptic potentials in the VS neurons, consisting of a mixture of EPSP's and IPSP's (Allen et al. 1972a, b; Bruggencate et al. 1972a, b, c; Andersson and Oscarsson 1978a). The IPSP's are determined by signals coming via the cerebellar cortex, i.e., via the axons of Purkinje cells, while the EPSP's are due to signals coming mainly via collaterals of the spino-cerebellar fibres. Ablation of the cerebellum results in the twofold increase of the resting discharge of VS neurons (Wylie and Felpel 1971; Orlovsky 1972b; Arshavsky et al. 1978c; see also Fig. 40, REST), as well as in the disappearance of the inhibitory components in the responses evoked by stimulation of the spinal cord or peripheral nerves (Whylie and Felpel 1971; Bruggencate et al. 1972c).

Fastigio-vestibular fibres from the rostral part of the fastigial nucleus project to the ipsilateral Deiters' nucleus, while the fibres from the caudal part project to the contralateral one (Brodal et al. 1962). Unlike the axons of Purkinje cells, the fastigial axons terminate mainly on smaller neurons. The fastigial neurons exert excitatory action upon VS neurons (Ito et al. 1970b).

Due to the powerful cerebello-vestibular connections, the cerebellum can strongly influence the activity of spinal neurons, the VS tract mediating these influences. Removal of the cerebellum causes extensor hypertonus (alpha-rigidity). The rigidity must be caused by the increased activity of VS neurons since it disappears after ablation of Deiters' nucleus (Batini et al. 1957). Stimulation of the vermis of the cerebellar anterior lobe results in inhibition of the decerebrate rigidity (Sherrington 1898; Dow and Moruzzi 1958), accompanied by hyperpolarization of the extensor motoneurons due to the cessation of the tonic excitatory influences (Terzuolo 1959; Llinás 1964). This phenomenon is, to a great extent, accounted for by the inhibitory action of Purkinje cells upon VS neurons.

b) Activity of VS Neurons During Locomotion and Scratching

The VS neurons in Deiters' nucleus, controlling the spinal hindlimb centre, were identified by their antidromic response to stimulation of the VS tract at the level of the first lumbar segment (Fig. 39 A–D). The activity of VS neurons during locomotion was considered in relation with the movements of the ipsilateral hindlimb; the scratch reflex was evoked on the side on which the neuron was recorded.

Let us first consider the activity of VS neurons in cats with intact cerebellum. An example of the discharge of a VS neuron during locomotion is shown in Fig. 39E (Orlovsky 1972b). During the latent period of locomotion the discharge frequency of the neuron increases. Such increase of activity before the beginning of locomotion was observed an in many other VS neurons as well. As stepping movements begin, the neuron's activity becomes rhythmical and related to limb movements: the discharge frequency increases during the swing phase, reaches the maximum by the end of this phase (i.e., by the onset of the extensor activity) and sharply decreases at the end of the stance phase after

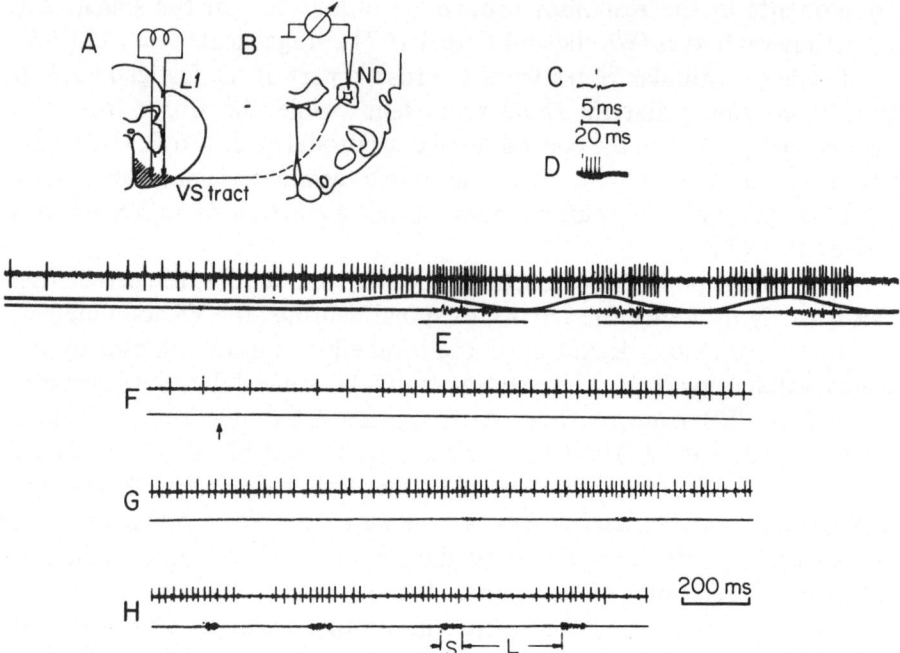

**Fig. 39A, B.** Recording and identification of VS neurons. Neurons were recorded from Deiters' nucleus (*ND*) and identified by the antidromic response to stimulation of the VS tract in the *L1* segment (*hatched area* in **A**). **C, D** Antidromic responses of a neuron to a single stimulus and to a train of impulses (400 pps) applied to the VS tract. **E–H** Activity of VS neurons during locomotion (**E**) and scratching (**F–H**). In **E**, the neuron is recorded together with the gastrocnemius EMG and the movement of ipsilateral hindlimb (protraction – up); stance phases of the limb are shown by *horizontal lines*. **F–H** Continuous recording of the activity of the VS neuron and the gastrocnemius EMG; the *arrow* indicates the beginning of stimulation of the C2 spinal segment (Orlovsky 1972b; Arshavsky et al. 1978c)

cessation of the extensor activity[7]. With more intense stepping, the rhythmical modulation of the VS neuron becomes more pronounced (cf. the first and third cycles in Fig. 39E).

---

[7] It is hard to estimate the relative amount of rhythmically active VS neurons (and neurons of other descending tracts), i.e., the units having modulation of the discharge related to the locomotor (or scratching) rhythm. Usually modulated neurons could be found only at the beginning of the experiment, when they might constitute the majority of neurons; later, in the course of the experiment, their relative amount decreased. In some experiments, modulated neurons were not found at all. The absence of rhythmical modulation or its disappearance in the course of the experiment seems to be accounted for by a poor functional state of the brain, especially of the cerebellum that plays a crucial role in modulating VS neurons (see below). Further we shall deal entirely with neurons having distinct discharge modulation

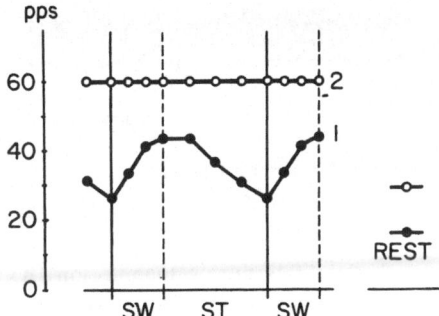

**Fig. 40.** Relationship between the discharge frequency of the "average" VS neuron and the phase of the step of the ipsilateral hindlimb during locomotion of cats with intact cerebellum (*1*) and decerebellate ones (*2*). For both cases, the average frequency of the background activity is also indicated (*REST*). Abbreviations as in Fig. 16 (Orlovsky 1972b)

Similar behaviour was observed in most rhythmically active neurons: they were mostly excited at the end of the swing phase or at the beginning of the stance phase, the discharge frequency reaching 100–200 pulses s$^{-1}$.

Figure 40 shows the frequency curve for the "average" VS neuron. The curve characterizes the overall impulse flow from Deiters' nucleus to the lumbo-sacral spinal cord in different phases of the step. The flow is maximum at the beginning of the stance phase, i.e., by the onset of the extensor activity, and then gradually decreases.

Similar results were obtained by Udo and his colleagues (Udo 1975; Udo et al. 1976): during locomotion of the thalamic cat, neurons of Deiters' nucleus were mostly excited in-phase with extensor activity.

The rhythmical activity of VS neurons during scratching is basically similar to that during locomotion, as one can conclude from the example in Fig. 39 F–H, from the phase distribution (Fig. 41A), and from the frequency curve for the "average" neuron (Fig. 41B) (Arshavsky et al. 1975b, 1978c).

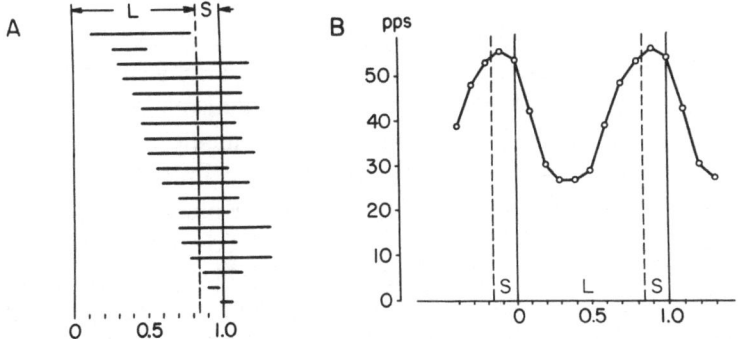

**Fig. 41A, B.** Relationship between the activity of VS neurons and the phase of the scratch cycle (actual scratching). **A** The phase distribution of 19 VS neurons. **B** The discharge frequency of the "average" VS neuron as a function of the phase of the cycle (Arshavsky et al. 1978c)

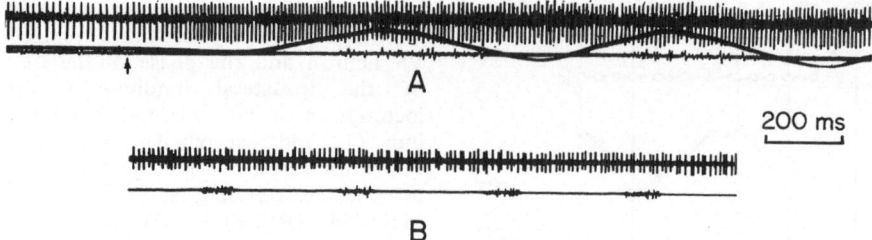

**A**

200 ms

**B**

**Fig. 42A, B.** Activity of VS neurons in decerebellate cats during locomotion (**A**) and scratching (**B**). In **A**, the *lower trace* is the gastrocnemius EMG, the *middle trace* shows the movement of the ipsilateral hindlimb (protraction – up). The *arrow* indicates the beginning of stimulation of the locomotor region. In **B**, the *lower trace* is the gastrocnemius EMG (Orlovsky 1972b; Arshavsky et al. 1978c).

Thus, in both movements, the overall activity of VS neurons increases at the end of the flexor phase, reaching the maximum by the beginning of the extensor phase.

The rhythmical modulation of the discharge of VS neurons during locomotion and scratching is performed by the cerebellum. In the decerebellate animals all rhythmical modulation was absent (Fig. 42). The only effect correlated with the appearance of rhythmical movements was an increase of the tonic discharge rate up to 60 pulses $s^{-1}$ (on the average) during locomotion (line 2 in Fig. 40) and 50 pulses $s^{-1}$ during scratching.

## 2. Reticulo-Spinal Tract

### a) General Characteristics

Most studies of the RS system emphasize its complex character. Various subdivisions of this system have different pharmacological characteristics and different effects upon spinal mechanisms.

Most RS fibres arise from the medial ponto-medullary reticular formation, mainly from the nuclei reticularis gigantocellularis, ventralis, pontis caudalis and pontis oralis (Brodal 1957, 1969; Rossi and Zanchetti 1957; Torvik and Brodal 1957; Magni and Willis 1963; Nyberg-Hansen 1966; Petras 1967; Ito et al. 1970a; Bezhenaru 1971; Kostyuk 1973; Pompeiano 1973; Peterson et al. 1974, 1975, 1979; Smirnov and Potechina 1974; Fukushima et al. 1979; Peterson 1979; Blessing et al. 1981). In the spinal cord the RS fibres descend in the medial part of the ventral funiculus (the medial RS tract) and in the lateral funiculus (the lateral RS tract). These two tracts originate from different though partly overlapping areas of the reticular formation. The medial RS tract arises primarily from the ipsilateral pontine nuclei and the rostro-dorsal part of the nucleus gigantocellularis. The lateral RS tract originates mainly from the ipsilateral (and partly contralateral) caudo-ventral part of the nucleus gigantocellularis and nucleus ventralis.

The conduction velocity of RS fibres ranges from 20–30 to 120–160 m s$^{-1}$ (Magni and Willis 1963; Wolstencroft 1964; Shapovalov et al. 1967; Grillner and Lund 1968; Orlovsky 1970a; Eccles et al. 1975a).

RS fibres reach the lumbo-sacral segments of the spinal cord terminating mostly in the laminae VII and VIII and, to a smaller extent, in the lamina XI ipsi- and partly contralaterally (Nyberg-Hansen 1965; Peterson et al. 1975). The RS pathways are not somatotopically organized (Torvic and Brodal 1957; Kuypers et al. 1962; Peterson et al. 1974; Eccles et al. 1976; Blessing et al. 1981). The majority of RS fibres sending terminal branches to the cervical enlargement also have axon branches extending to lower spinal levels – thoracal and especially lumbar ones (Peterson et al. 1975).

The RS neurons have low background activity (1–20 pulses s$^{-1}$) (Orlovsky 1970a; Smirnov and Potechina 1974; Eccles et al. 1975a). In this respect they differ from other reticular neurons which usually fire at rest with frequencies of 30–100 pulses s$^{-1}$ (Moruzzi 1954; Rossi and Zanchetti 1957; Orlovsky 1970a; Pompeiano 1973).

The RS system exerts highly different influences upon spinal mechanisms. Thus, experiments with stimulation of the reticular formation were sometimes giving contradictory results. Magoun and colleagues (Magoun and Rhines 1946; Rhines and Magoun 1946; Niemer and Magoun 1947) concluded that the reticular formation exerts global influences upon spinal mechanisms. They described two areas in the reticular formation, whose stimulation results in either diffuse facilitation or diffuse inhibition of the spinal reflexes and muscle tone of all four limbs. However, later it was found that the reticular formation can exert rather specific, reciprocal influences upon reflexes of the antagonistic muscles (Sprague and Chambers 1953, 1954; Gernandt and Thulin 1955). These data were confirmed in experiments with intracellular recordings. In accordance with earlier findings, stimulation of the caudo-ventral part of the nucleus gigantocellularis was found to produce long latency and slowly rising hyperpolarization of both flexor and extensor motoneurons (Llinás and Terzuolo 1964, 1965; Jankowska et al. 1968; Kostyuk 1973). This hyperpolarization is produced mainly by slow conducting RS fibres descending in the lateral funiculus (Ito et al. 1970a; Peterson et al. 1975, 1978b).

A different effect is produced by stimulation of the pontine nuclei and the rostro-dorsal part of the nucleus gigantocellularis, as well as of the medial longitudinal fasciculus originating in the pontine reticular formation (Shapovalov et al. 1967; Grillner and Lund 1968; Grillner et al. 1968a, b, 1971; Kostyuk 1973; Shapovalov 1975a, b). Stimulation of these structures evokes short latency (sometimes, monosynaptic) EPSP's in flexor motoneurons and di- and polysynaptic IPSP's in extensor ones. The RS tract excites both alpha- and static gamma-motoneurons of flexors (Grillner et al. 1966b, 1969; Bergmans and Grillner 1968; Grillner 1969). The reciprocal influences upon the motoneurons are produced by fast conducting fibres of the medial RS tract. However, other authors (Wilson and Yoshida 1969; Peterson et al. 1978b, 1979; Peterson 1979) found that stimulation of the rostral reticular formation evokes monosynaptic EPSP's not only in flexor, but also in extensor motoneurons.

Afferent connections of the RS neurons are extremely diverse. The neurons receive signals from the somatic and vestibular receptors, cerebral cortex, cerebellum, tectum, and some other structures. We shall confine ourselves to a description of the spinal and cerebellar inputs. The spino-reticular fibres terminating in the medial ponto-medullary reticular formation ascend in the lateral funiculus (Brodal 1957; Rossi and Zanchetti 1957; Mehler et al. 1960; Bowsher and Westman 1970; Pompeiano 1973). Through this pathway the RS neurons receive signals mainly from the flexor reflex afferents and, to a smaller extent, from Golgi tendon organs. These signals exert both excitatory and inhibitory influences upon RS neurons. The receptive fields of RS neurons are extremely large. Most of the neurons respond to stimulation of cutaneous and muscle receptors of more than one limb (Magni and Willis 1964a, b; Wolstencroft 1964; Udo and Mano 1970; Peterson and Felpel 1971; Pompeiano and Barnes 1971; Peterson et al. 1974; Eccles et al. 1975a).

The cerebello-reticular pathway originates mainly from the rostral and, to a smaller extent, from the caudal parts of the fastigial nucleus (Jansen and Brodal 1954; Brodal 1957; Walberg et al. 1962; Batton et al. 1977). It terminates in the medial reticular formation (mostly contra-laterally) (Walberg et al. 1962; Eccles et al. 1975a; Batton et al. 1977). Fastigio-reticular connections are not somatotopically organized. Stimulation of the fastigial nucleus evokes mono- and polysynaptic EPSP's in RS neurons (Ito et al. 1970b; Eccles et al. 1975a). Comparing reactions of the RS and fastigial neurons to peripheral stimuli, Eccles and colleagues concluded that at least some components of the reactions of RS neurons are determined by cerebellar input (Eccles et al. 1975a). After cerebellar ablation, the resting discharge of RS neurons becomes weaker, probably, because of the cessation of the excitatory inflow from the fastigial nucleus (REST in Fig. 44) (Orlovsky 1970c; Orlovsky and Pavlova 1972a; Pavlova 1977).

## b) Activity of RS Neurons During Locomotion and Scratching

RS neurons were recorded from the medial reticular formation of the rostral part of the medulla and of the pons; they were identified by the antidromic response to stimulation of the ventral funiculus of the spinal cord near the midline at the L1 level (Fig. 43 A–D) (Orlovsky 1970b, c; Pavlova 1977). Thus the neurons giving rise to the medial reticulo-spinal tract seem to have been mainly involved. Since

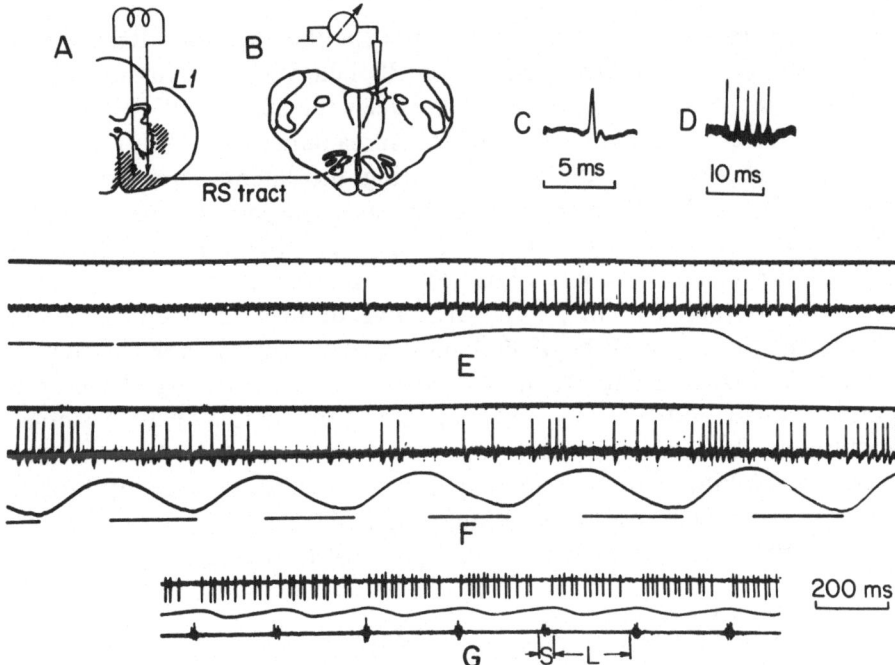

Fig. 43A, B. Recording and identification of RS neurons. Neurons were recorded from the medial reticular formation of the pons and medulla oblongata and identified by the antidromic response to stimulation of the RS tract in the *L1* segment (*hatched area* in A). C, D Antidromic responses of a neuron to a single stimulus and to a train of impulses (500 pps) applied to the RS tract. E, F Continuous recording of the activity of a RS neuron during locomotion. *Upper trace,* stimulation of the locomotor region; *lower trace,* movement of the ipsilateral hindlimb (protraction – up). Stance phases of the limb are shown by *horizontal lines.* G Activity of a RS neuron during scratching; the *lower trace* is the gastrocnemius EMG (Orlovsky 1970b; Pavlova 1977)

most fibres of this tract do not cross the midline, the activity of RS neurons during locomotion was compared with the activity of the ipsilateral hindlimb muscles; the scratch reflex was evoked on the side on which the neuron was recorded.

Let us first consider the activity of RS neurons during locomotion in the cat with intact cerebellum. In the overwhelming majority of RS neurons, the tonic excitation preceded the beginning of stepping movements (which could arise either spontaneously or as a response to stimulation of the locomotor region). The mean discharge frequency of RS neurons during locomotion is 10–15 times higher than at rest. During locomotion, a part of the neurons exhibits, besides tonic activity, a rhythmical one, i.e., modulation of the discharge frequency with locomotor rhythm. An example of a "modulated" RS neuron is shown in Fig. 43E, F. The neuron becomes active prior to the beginning of stepping, and then fires mainly at the end of the stance phase and in the swing phase.

Figure 44 shows the frequency curve for the "average" RS neuron. It is seen that the overall activity of RS neurons is maximum in the swing phase. A similar relationship between the activity of RS neurons and the phase of the step was found by Shimamura et al. (1982).

The behaviour of RS neurons during the latent period of scratching differs strikingly from that during the latent period of locomotion. Stimulation of the pinna either does not influence or suppresses the activity of RS neurons. Thus, only a few neurons (less than 10%) are active during scratching. The discharge of these neurons is usually modulated in the rhythm of limb movements. An example of such a neuron is shown in Fig. 43G. The neuron is active during almost the whole L-phase and is silent in the S-phase of the scratch cycle. A similar pattern of rhythmical activity was also found in other neurons that were not inhibited during scratching. Thus, the patterns of the rhythmical activity of RS neurons during locomotion and scratching are similar: the neurons are more excited in the flexor phase of the cycle.

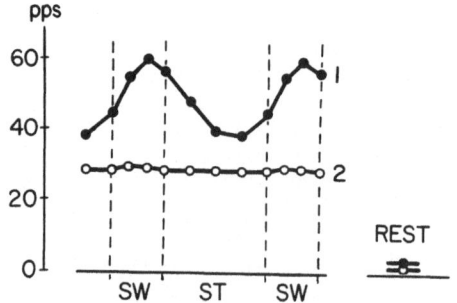

Fig. 44. Relationship between the discharge frequency of the "average" RS neuron and the phase of the step of the ipsilateral hindlimb during locomotion of cats with intact cerebellum (1) and decerebellate ones (2). For both cases, the average frequency of the background activity is also indicated (REST). Abbreviations as in Fig. 16 (Orlovsky 1970b)

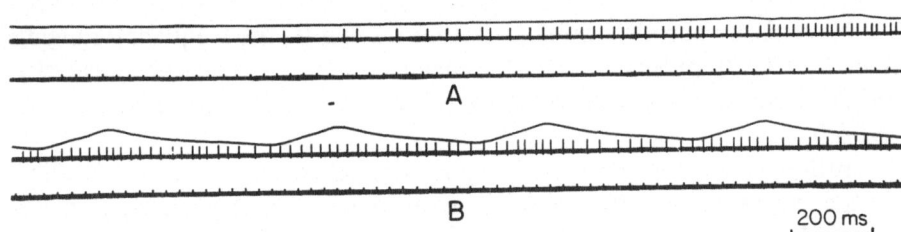

**Fig. 45A, B.** Activity of a RS neuron during locomotion of the decerebellate cat (**B** is the continuation of **A**). The *upper trace* is the movement of the ipsilateral hindlimb (protraction — up); the *lower trace,* stimulation of the locomotor region (Orlovsky 1970c)

The cerebellum is responsible for the rhythmical modulation of the RS neuron discharge, at least during locomotion. In decerebellate cats, most RS neurons exhibit only tonic activity during locomotion (Fig. 45); the mean frequency increases from 2 pulses $s^{-1}$ (REST) to 30 pulses $s^{-1}$ (curve 2 in Fig. 44). Only a few neurons had weak rhythmical activity. During scratching in decerebellate cats all the recorded RS neurons were inhibited.

## 3. Rubro-Spinal Tract

### a) General Characteristics

A detailed description of structural and functional characteristics of the RbS tract can be found in a number of books and reviews (Pompeiano and Brodal 1957b; Massion 1967; Brodal 1969; Hongo et al. 1969a, b; Kostyuk 1973; Fanardjian 1975; Shapovalov 1975a, b).

The RbS tract originates mainly in the caudal part of the red nucleus (Pompeiano and Brodal 1957b) which is traditionally called the "magnocellular part", though it contains neurons of different size. The RbS fibres cross the midline and descend in the lateral funiculus down to the lumbo-sacral segments. They terminate mostly in the intermediate zone of the spinal cord (the laminae V—VII). The RbS tract is somatotopically organized. The fibres terminating in the cervical and lumbo-sacral segments of the spinal cord originate mostly from the dorso-medial and ventro-lateral parts of the red nucleus, respectively (Pompeiano and Brodal 1957b; Nyberg-Hansen and Brodal 1964; Padel et al. 1972; Eccles et al. 1975b; Larsen and Yumiya 1980). The conduction velocity of RbS fibres ranges from 30 to 120—150 m $s^{-1}$ (Tsukahara et al. 1964, 1967; Fanardjian and Sarkisjian 1969; Kostyuk and Pilyavsky 1969; Bayev and Kostyuk 1972; Padel et al. 1972; Eccles et al. 1975c).

Most RbS neurons exhibit a resting discharge, the frequency ranging from 2–5 to 30 pulses $s^{-1}$, with the mean value of 10 pulses $s^{-1}$ (Orlovsky 1972c; Orlovsky, Pavlova 1972a; Eccles et al. 1975c; Arshavsky et al. 1978e).

Influence of the RbS tract upon the spinal neurons was studied in the experiments with stimulation of the red nucleus. This stimulation induces a flexion of the contralateral limbs (the hind- or forelimb depending on the site of stimulation) and facilitates flexion reflexes (Sherrington 1898; Pompeiano 1959, 1967; Sasaki et al. 1960; Thulin 1963; Massion 1967). Stimulation of the red nucleus evokes polysynaptic and in some cases monosynaptic EPSP's mainly in flexor alpha-motoneurons (Sasaki et al. 1960; Shapovalov and Shapovalova 1966; Kostyuk and Pilyavsky 1969; Hongo et al. 1969a, b; Kostyuk 1973; Shapovalov 1975a, b). Besides, the RbS tract excites flexor gamma-motoneurons (Appelberg 1962; Appelberg, Kosary 1963; Appelberg et al. 1975).

Though a predominant influence of the red nucleus upon flexor motoneurons is evident, some authors have reported that the RbS tract can excite extensor motoneurons as well (Burke et al. 1970; Kostyuk 1973; Ghez 1975; Shapovalov 1975a, b; Giuffrida et al. 1980). This question was most thoroughly investigated by Ghez (1975), who demonstrated that stimulation of small zones within the red nucleus is capable of eliciting contraction of both flexor and extensor muscles. However, "the flexor zones" are more numerous, especially for the hindlimb muscles.

Afferent signals reach the red nucleus via two main input systems, i.e., cerebro-rubral and cerebello-rubral ones. Here, we shall deal only with the latter. The caudal part of the red nucleus, giving rise to the RbS tract, receives signals from the contralateral interpositus nucleus (IN) which, in turn, is under the control of the pars intermedia (Jansen and Brodal 1954; Voogd 1964; Angaut and Bowsher 1965; Courville 1966, 1968; Eccles et al. 1967; Anderson 1971; Nakamura and Mizuno 1971; Flumerfelt 1978; Nakamura et al. 1978). Cerebello-rubral fibres terminate on the cell somas and proximal dendrites of the RbS neurons (unlike the cerebro-rubral fibres which terminate on the distal dendrites). About 50 interpositus axons converge upon each RbS neuron (Toyama et al. 1970).

Cerebellar nuclei exert excitatory influences upon all the target structures. In particular, stimulation of the IN evokes monosynaptic EPSP's in RbS neurons (Tsukahara et al. 1964, 1967; Eccles et al. 1967; Toyama et al. 1968, 1970; Fanardjian and Sarkisjian 1969; Anderson 1971; Allen and Tsukahara 1974). After ablation of the cerebellum, the activity of RbS neurons is considerably reduced (REST in Fig. 47) (Massion 1967; Orlovsky 1972c; Arshavsky et al. 1978e) due to the cessation of tonic excitatory inflow from the IN (Tsukahara et al. 1965). Besides, cerebellotomy leads to a sharp decrease of responses of RbS neurons to stimulation of peripheral nerves or somatic receptors (Massion 1967; Nishioka and Nakahama 1973; Eccles et al. 1975c; Fanardjian and Manvelian 1976). This means that peripheral signals reach RbS neurons mainly through the cerebellum. After cerebellar ablation only a minor component of the response persists (Massion 1967; Eccles et al. 1975c), which seems to be mediated by direct spino-rubral fibres.

The RbS tract mediates influences of the intermedial zone of the cerebellum upon the spinal cord. Stimulation of the IN evokes short latency EPSP's in the motoneurons of distal muscles (Shapovalov et al. 1972) and motor responses (more commonly, of the flexor type) of the ipsilateral limbs (Pompeiano 1959, 1967; Asanuma and Hunsperger 1975; Giuffrida et al. 1980). Cooling of the IN, in chronic experiments on monkeys, results in flexor hypometry in the ipsilateral limbs (Uno et al. 1973). In contrast, removal or cooling of the pars intermedia of the cerebellar cortex (which inhibits the IN neurons) leads to an increase of the tonic activity of flexor muscles (Yu 1972; Yu et al. 1973) as well as to an increase of the amplitude and duration of the flexor phase of the step during locomotion (Udo et al. 1979a, b, 1980). All these effects are mediated by the RbS tract.

## b) Activity of RbS Neurons During Locomotion and Scratching

Since the RbS tract crosses the midline, activity of RbS neurons was considered together with movements of the contralateral hindlimb, while the scratch reflex was evoked on the side contralateral to the neuron. Neurons were identified by the antidromic response to stimulation of the RbS tract (Fig. 46 A–D).

Let us first consider the experiments on cats with intact cerebellum. Prior to the beginning of locomotion, most RbS neurons increase their discharge rate, while during locomotion they exhibit rhythmical activity related with limb movements (Fig. 46E) (Orlovsky 1972c). As in neurons of other descending tracts, the rhythmical activity of RbS neurons was more pronounced during more intense locomotion. The majority of neurons are mostly excited in the swing (flexor) phase of the step, though in some neurons maximum activity is observed in the stance phase.

Figure 47 shows the frequency curve of the "average" RbS neuron. One can see that the overall flow of impulses in the RbS tract is maximum in the swing phase.

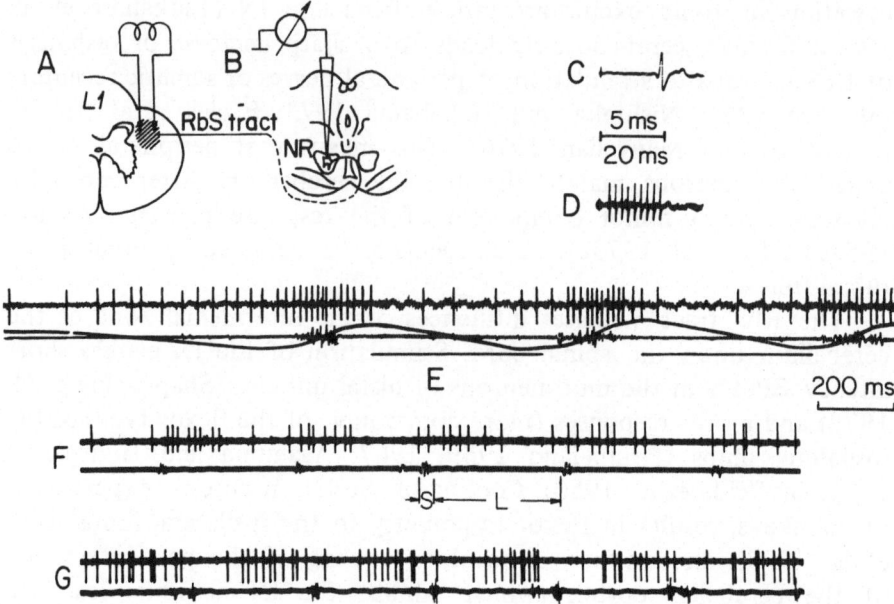

**Fig. 46A, B.** Recording and identification of RbS neurons. Neurons were recorded from the red nucleus (*NR*) and identified by the antidromic response to stimulation of the RbS tract on the contralateral side of the spinal cord in the *L1* segment (*hatched area* in **A**). **C, D** Antidromic responses of a neuron to a single stimulus and to a train of impulses (500 pps) applied to the RbS tract. **E** Activity of a RbS neuron during locomotion recorded simultaneously with the EMG of m.tibialis anterior and the movement of the contralateral hindlimb (protraction — up); stance phases of the limb are shown by *horizontal lines*. **F, G** Activity of two RbS neurons during scratching; the *lower trace* is the gastrocnemius EMG (Orlovsky 1972c; Arshavsky et al. 1978e)

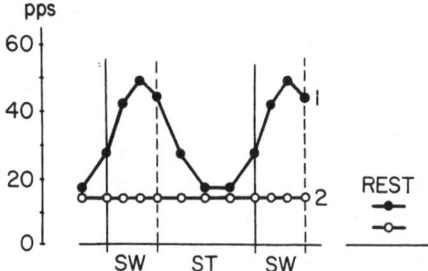

**Fig. 47.** Relationship between the discharge frequency of the "average" RbS neuron and the phase of the step of the contralateral hindlimb during locomotion of cats with intact cerebellum (*1*) and of decerebellate ones (*2*). For both cases, the average frequency of the background activity is also indicated (*REST*). Abbreviations as in Fig. 16 (Orlovsky 1972c)

Corresponding results were obtained while studying the activity of RbS neurons during scratching (Arshavsky et al. 1978e). Usually the neurons get tonically active during the latent period of rhythmical scratching, then rhythmical modulation of the discharge arises in many neurons. One can see from the examples given in Fig. 46F, G and from the phase distribution (Fig. 48A) that RbS neurons discharge in various parts of the cycle. Nevertheless, the frequency curve for the "average" RbS neuron (Fig. 48B) shows that the overall activity of the population of RbS neurons is higher in the flexor (L) phase than in the extensor (S) phase, as in locomotion (Fig. 47). The discharge frequency of RbS neurons during scratching and locomotion usually ranges from 30 to 100 pulses $s^{-1}$.

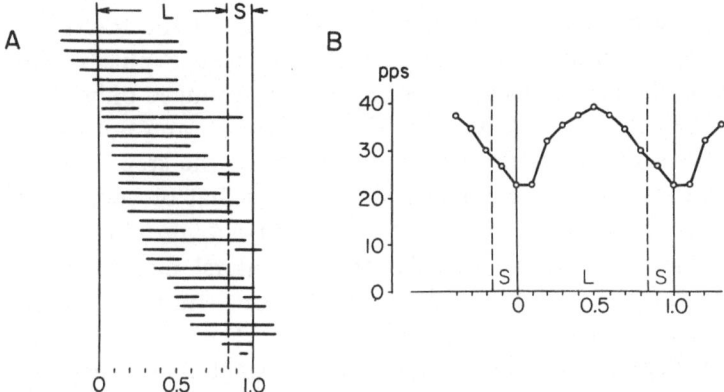

Fig. 48A, B. Relationship between the activity of RbS neurons and the phase of the scratch cycle (actual scratching). A The phase distribution of 35 RbS neurons. B The discharge frequency of the "average" RbS neuron as a function of the phase of the cycle (Arshavsky et al. 1978e)

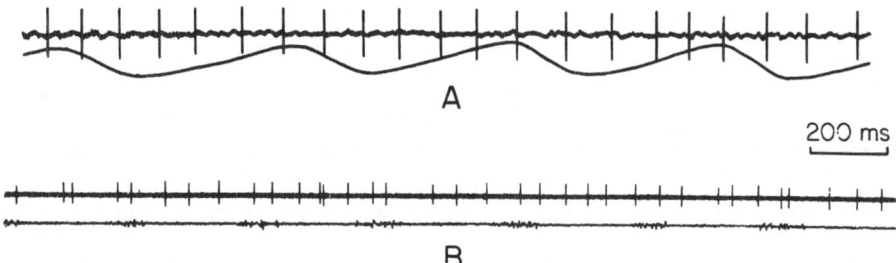

Fig. 49A, B. Activity of RbS neurons in decerebellate cats during locomotion (A) and scratching (B). In A, the *lower trace* is the movement of the contralateral hindlimb (protraction − up). In B, the *lower trace* is the gastrocnemius EMG (Orlovsky 1972c; Arshavsky et al. 1978e)

The rhythmical activity of RbS neurons (like that of VS and RS neurons) is determined by the cerebellum. In decerebellate animals, during both locomotion and scratching, one can observe a tonic increase of the discharge rate while, in most neurons, any rhythmical modulation of the discharge is absent (Fig. 49). The mean value of the discharge frequency during movements is 14 pulses $s^{-1}$ (line 2 in Fig. 47).

## 4. Influences of Descending Tracts upon Motor Activity

The results presented previously show that provided the cerebellum is intact, rhythmical signals related to motor activity are conveyed from the brain stem to the spinal cord through the descending tracts. To estimate their role in the activity of spinal mechanisms, the following experiments were carried out (Orlovsky 1972a). Each of the tracts was stimulated during locomotion by short trains of electric pulses. The strength of the applied stimulus was so weak that it produced only slight motor effects when the cat was inactive. The train duration was comparable with durations of the stance or swing phases of the step; the frequency of the pulses was approximately the same as that of neurons of descending tracts in the phase of their maximum activity. Figure 50 shows the effects produced by stimulation of Deiters' nucleus upon the antagonistic muscles of the ankle joint (flexor — m.tibialis anterior and extensor — m.gastrocnemius lateralis). The first stimulus was applied in the stance phase, i.e., when extensors (including m.gastrocnemius) are active; it resulted in considerable increase of the extensor activity. The second stimulus was applied in the swing phase and at the very beginning of the stance phase; and again it resulted in considerable increase of the extensor activity, but did not affect the flexor one.

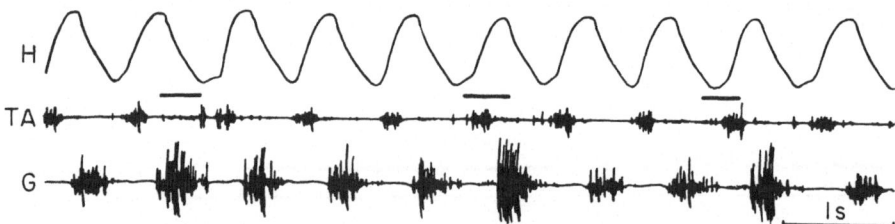

**Fig. 50.** Effects of stimulation of Deiters' nucleus druing locomotion upon the muscle activities of the ipsilateral hindlimb. Stimuli (pulses 0.1 ms, 50 $s^{-1}$, 150 $\mu$A) were applied during periods marked by *horizontal lines*. Movement of the hip (*H*, flexion — up) and EMG's of m.tibialis anterior (*TA*) and m.gastrocnemius (*G*) are recorded (Orlovsky 1972a)

Finally, Deiters' nucleus was stimulated at the very end of the stance phase and in the swing phase. The flexor activity did not change, and no additional activity of the- extensor was observed in the swing phase, but the subsequent extensor burst in the stance phase markedly increased. Similar effects of Deiters' nucleus stimulation were observed in the hip and knee extensors: additional extensor excitation was maximum provided the stimulus was applied at the end of the swing phase and in the stance phase, i.e., prior or during the locomotor extensor burst. If the cat had not been fixed in an animal holder, the increase of the propulsive force due to the stimulation would have accelerated its running.

Figure 51 shows the effects produced by stimulation of the medial reticular formation upon the same muscles. The stimulus applied at the beginning of the swing phase (A) sharply increased the flexor activity; besides, the extensor activity in the subsequent (stance) phase was also somewhat increased. Stimulation performed at the end of the swing phase (B) slightly increased the flexor burst and decreased the extensor one, but did not influence the timing of the step cycle: the flexor burst terminated at the right moment and was followed by the extensor burst. The stimuli applied either at the beginning (C) or in the middle (D) of the stance phase did not affect the flexor, but suppressed the

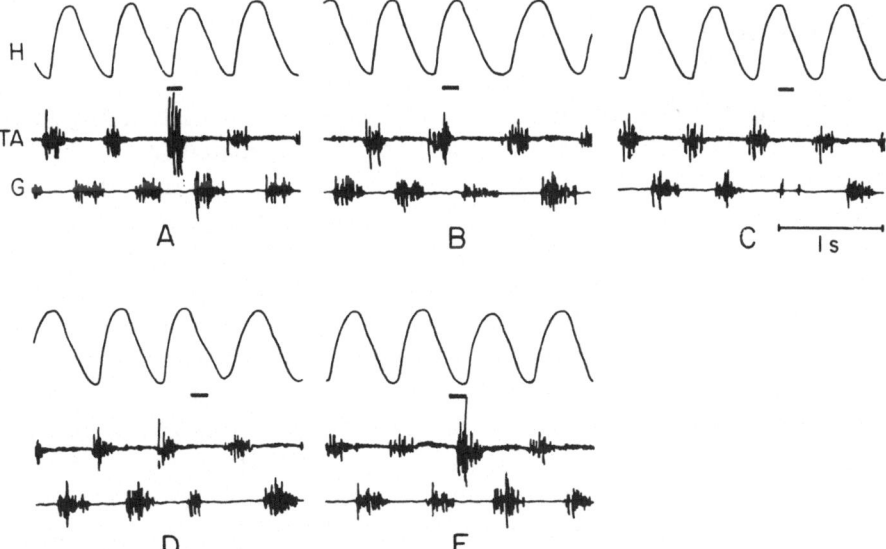

Fig. 51 A—E. Effects of stimulation of the medial reticular formation of the medulla oblongata upon the activity of muscles of the ipsilateral hindlimb during locomotion. Stimuli (pulses 0.1 ms, 100 s$^{-1}$, 120 $\mu$A) were applied during periods marked by *horizontal lines*. Abbreviations as in Fig. 50 (Orlovsky 1972a)

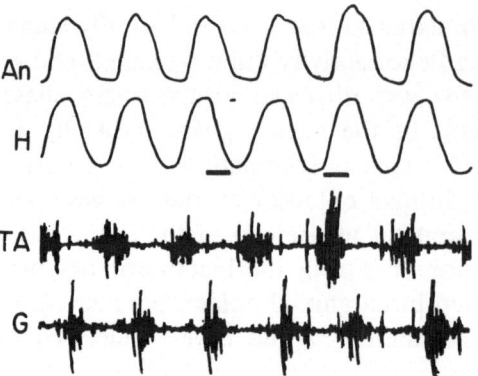

**Fig. 52.** Effects of stimulation of the red nucleus during locomotion upon the joint movements and the activity of muscles of the contralateral hindlimb. Stimuli (pulses 1.0 ms, 50 s$^{-1}$, 100 μA) were applied during periods marked by *horizontal lines*. *An:* movement at the ankle joint (flexion – up). Other abbreviations as in Fig. 50 (Orlovsky 1972a)

extensor activity. Finally, stimulation performed at the end of the stance phase (E) led to the increased flexor burst.

Figure 52 shows the effects of stimulation of the red nucleus. The stimulus applied in the stance phase was not effective, but applied in the swing phase, it facilitated the flexor activity that resulted in the increase of the joint flexion.

Thus, moderate stimulation of the descending tracts causes considerable changes in muscle activity, but does not affect the temporal pattern of stepping, i.e., the phases of muscular activity in the cycle.

## 5. Conclusion

The data presented in this chapter have shown that during locomotion and scratching the flows of impulses modulated in the rhythm of limb movements are conveyed by descending tracts from the brain stem to the spinal cord. The crucial role in organizing the rhythmical activity of neurons giving rise to the descending tracts is played by the cerebellum: after cerebellar ablation the rhythmical activity of VS, RS and RbS neurons ceases. This result is quite striking since it demonstrates that these neurons can receive signals from the spinal cord not only through the cerebellum. The RS neurons receive mono- and polysynaptic influences from the spinal cord through the spino-reticular fibres; VS neurons – through the collaterals of the DSCT fibres (and, perhaps, through spino-vestibular fibres); finally, some spinal fibres terminate on RbS neurons. But all these inputs seem to be inefficient in producing the rhythmical modulation of the activity of VS, RS and RbS neurons. Probably, these inputs participate in tonic activation of the neurons which is observed during locomotion and scratching both in animals with intact cerebellum and in decerebellate ones.

The pattern of rhythmical signals conveyed by descending tracts makes good sense in terms of function. The maximum flow of impulses conveyed by a given tract coincides_with the phase of maximum activity of those muscles on which the tract exerts the greatest excitatory effect: the VS tract is mostly active in-phase with extensors, while the RS and RbS ones — in-phase with flexors.

It should be noted that the coincidence between the activities of descending tracts and corresponding muscle groups is far from being complete. For example, the overall activity of RbS neurons during locomotion is the highest in the second half of the swing phase (Fig. 47) when activity of some flexors is already decreasing (see also Perret 1973, 1976). Besides, some RbS neurons have the firing pattern opposite to that obtained in the majority of the neurons: they fire in the extensor phase of the cycle (see also Fig. 48A for scratching).

Various patterns of the activity of neurons of the red nucleus were also observed in chronic experiments on cats trained to perform various forelimb movements (flexion and extension at the elbow joint, placing reaction). It was found that the activity of many neurons was correlated with movements, some units being active simultaneously with flexors, others with extensors (Ghez and Kubota 1977; Burton and Onoda 1978; Padel and Steinberg 1978). Nonuniform behaviour of these neurons might be accounted for by the complexity of the effects produced by the RbS tract in the spinal cord. As it has been mentioned above, this tract excites not only the flexor motoneurons but also, to some extent, the extensor ones (Ghez 1975). Besides, RbS fibres influence not only motoneurons, but also spinal interneurons from various reflex pathways (Hongo et al. 1969a, b). Phases of activity of spinal interneurons in the cycle can be extremely diverse (see, for example, the phase distribution for scratching, Fig. 11), and activity of "nontypical" RbS neurons might correlate with the activity of certain interneurons. Similar reservations can be made also for the VS tract and, especially, for the RS one, whose targets in the spinal cord are not fully known.

The experiments with stimulation of descending tracts during locomotion have revealed the "final" effects produced by descending commands. Like in nonactive animals (decerebrated or anaesthetized) stimulation of the VS tract during locomotion enhances the extensor activity; stimulation of the RS tract enhances the flexor activity and inhibits the extensor activity; finally, stimulation of the RbS tract enhances the flexor activity. However, all these stimulations causing considerable changes of the level of muscle activity do not influence the timing of this activity in the step cycle.

A different result was obtained while stimulating Deiters' nucleus and the RS tract during fictitious locomotion (Russel and Zajac 1979). It was found that in immobilized animals activation of descending tracts affects not only the ENG amplitudes, but also durations of the flexor and extensor phases of the step and, consequently, the locomotor rhythm. The different effects of stimulation observed during actual and fictitious locomotion  can be explained by the different origin of the locomotor rhythm in these two cases. During fictitious locomotion, the rhythm is determined exclusively by the central pattern generator. The regime of its activity can be easily changed even by very weak afferent influences, for example, by touching the toes of the hindlimb (Orlovsky and Feldman 1972a). During actual locomotion, the principle of rhythm generation is quite different: transition from one phase of the cycle to the next takes place provided the limb reaches a certain position (see Chap. I). In this case, the rhythm generator is much more independent of external influences. The influences have to be so powerful that they might either (1) accelerate  (or decelerate) the limb movement so that the limb reaches the critical posture (where the next phase starts) earlier or later than in the normal cycles; or (2) overcome the influences of afferent signals coming from the moving limb. Thus, in actual locomotion, the rhythmical commands transmitted by the VS, RS and RbS tracts exert practically no influences upon the temporal pattern of movements, but are rather "addressed" to the "outputs" of the pattern generator. Therefore, since the cerebellum may affect the spinal mechanisms only via the descending tracts, this result suggests that it can only increase or decrease the level of the muscle activity and not the phases of these activities in the step cycle.

A study of the effects of descending signals during scratching was not carried out. Nevertheless, the cerebellar influences upon the spinal cord in this case seem to be potentially the same as during locomotion; Indeed, the central spinal generator for scratching is very stable against external influences: neither deafferentation of the limb nor spinalization of the animal affects its temporal pattern (Sherrrington 1906b; 1910b; Deliagina et al. 1975).

The experimental results described in the present chapter show that the mean level of activity of VS and RbS neurons during both loco-motion and scratching is considerably higher than at rest. The behaviour of RS neurons in these two movements is different: during locomotion they are strongly activated, while during scratching most of them are silent. How can this difference be accounted for? As has been already mentioned in Chap. 1, in the animals deprived of the forebrain, locomotion can be evoked by stimulation of the locomotor region of the midbrain or subthalamus. According to morphological and electrophysiological data, these structures have direct (monosynaptic) connections with RS neurons (Rossi and Brodal 1956; Magni and Willis 1964a, b; Orlovsky 1970a). While locomotion is being evoked, one can observe a direct correlation between activation of RS neurons and the level of motor activity: as stimulation of the locomotor region becomes stronger, the latent periods both of activation of RS neurons and of the appearance of movements are shortened; higher activity of RS neurons corresponds to more intense locomotion (Orlovsky 1970b). On the basis of these data a hypothesis has been advanced according to which the RS tract originating from the reticular formation of the pons and medulla oblongata triggers (or, at least, participates in triggering) the spinal mechanisms of stepping (Orlovsky 1970b; see also Steeves and Jordan 1980; Eidelberg et al. 1981). If this hypothesis is true, it easily explains why the RS tract is inactive during scratching. Sherrington (1906a) found locomotion and scratching to be mutually exclusive. In our experiments it has been also observed that stepping movements, which spontaneously appeared in thalamic cats, would always cease whenever the scratch reflex was evoked. It seems likely that with activation of the mechanisms responsible for scratching, the activity of the RS system responsible for locomotion is suppressed.

The essentially different behaviours of RS neurons during locomotion and scratching are not determined by the cerebellum. Indeed, a strong excitation of RS neurons with elicitation of locomotion is observed in decerebellate cats as well. This excitation seems to be carried by fibres coming to RS neurons from the locomotor regions of the brain stem (Orlovsky 1970b). In Fig. 1, all the extracerebellar inputs to the descending tract neurons were called the "external" ones. In RS neurons one can see that the signals conveyed by the descending tract depend on both the signals coming from the cerebellum and those coming through the "external inputs" from other parts of the brain. We shall return to this problem in Chap. VI.

# IV Role of Different Input Signals for Generating Cerebellar Output Signals

## 1. Role of Signals Concerning Peripheral and Central Processes During Locomotion

The results described in the previous Chapter have shown that during locomotion and scratching, the cerebellum performs rhythmical modulation of the discharge frequency of neurons of descending tracts, modulation related to limb movements. The source of this rhythmical activity is not the cerebellum itself, but the rhythmical signals which are conveyed to the cerebellum from the spinal cord by the spino-cerebellar pathways. This is clearly seen from the following experiment. When during locomotion the treadmill band is abruptly stopped, both the stepping movements and the rhythmical activity in descending tracts immediately cease (Fig. 53A) irrespective of the continued stimulation of the locomotor region (Orlovsky 1970b). Besides, both the arrest (Fig. 53B) and acceleration (Fig. 53C) of the contralateral hindlimb

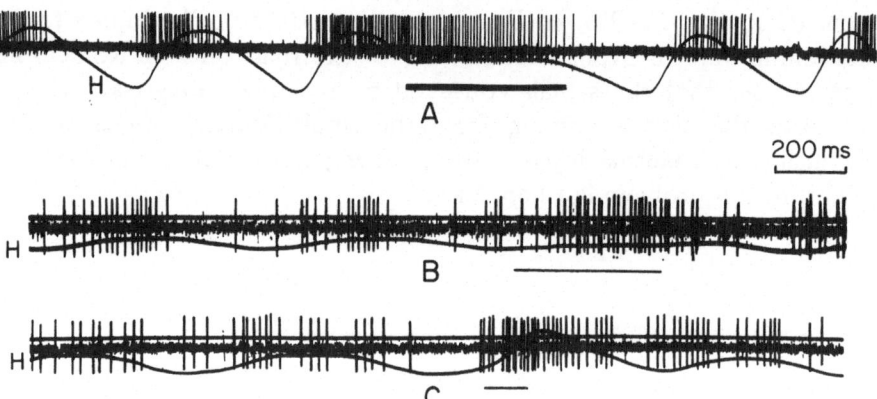

Fig. 53 A–C. Effects of disturbances of the hindlimb movements at the hip joint upon the activity of neurons of descending tracts during locomotion. **A** Influence of the delay of movement of the ipsilateral limb upon the activity of a RS neuron. **B, C** Influences of the delay (**B**) and acceleration (**C**) of movements of the contralateral limb upon the activity of a RbS neuron. Neurons were recorded together with the hip movement (*H*, flexion − up). Periods of the external action upon the limb are marked by *horizontal lines* (Orlovsky 1970b, 1972c)

evoke corresponding changes in the activity of neurons of the RbS tract. Thus, rhythmical activity in descending tracts depends on the processes taking place at the spinal level.

However, not every process or event that takes place at this level affects the activity of neurons of descending tracts. Figure 54 shows activities of three RbS neurons during locomotion. In A, flexion at the ankle joint in the swing phase was stopped by force, while in B it was accelerated. One can see that the disturbances of movements at the ankle joint do not affect the pattern of activity of RbS neurons. There was also no effect produced by extreme forced flexion at the knee joint (C). These experiments (Orlovsky 1972c) have clearly shown that disturbances of movements at the distal joints (knee and ankle) do not affect (or scarcely affect) the rhythmical activity of RbS neurons. In contrast, disturbances of the hip movement immediately affect the activity of RbS neurons. No doubt, all the disturbances of movements, both at distal and at proximal joints, are reflected in the afferent signals coming from the limb receptors. The question arises, why are the signals on disturbances of movements at the knee and ankle joints of no importance for rhythmical modulation of RbS neurons, while those on disturbances of the hip movements are very important? To explain these findings, a

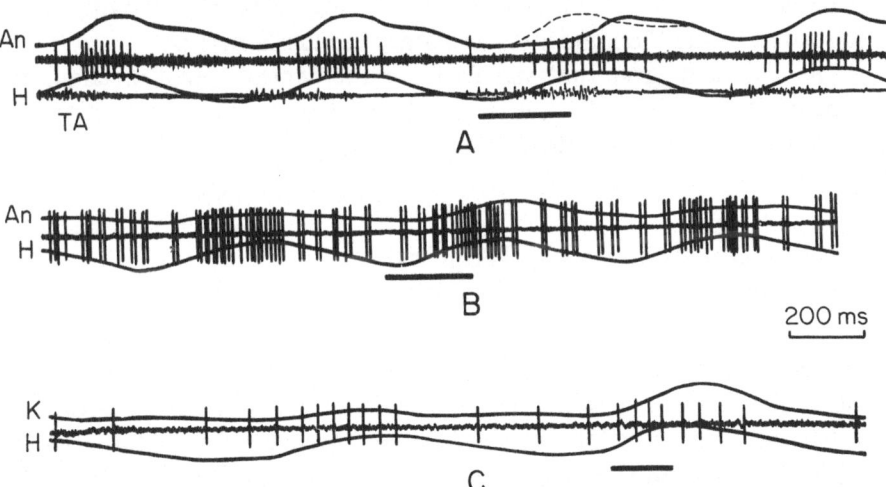

Fig. 54 A–C. Disturbances of the hindlimb movements at the ankle and knee joints produce no effects on the activity of RbS neurons in the contralateral red nucleus. A Delay of the ankle joint flexion produced by electrical stimulation of m.gastrocnemius. B Acceleration of the ankle joint flexion produced by electrical stimulation of m.tibialis anterior. C Acceleration of the knee flexion produced by an external force applied to the foot. Activity of neurons is recorded together with the hip angle (*H*) and ankle angle (*An*) or knee angle (*K*). In all the records, flexion up. In A, the presumed course of the *An* curve for an undisturbed cycle is shown by the *interrupted line; TA,* EMG of m.tibialis anterior (Orlovsky 1972c)

hypothesis has been advanced (Orlovsky 1972c) that the effect of disturbances of limb movements depends on the degree of their influence upon the central spinal mechanisms. Any disturbance of the hip movements evokes corresponding changes in the movements of other joints (Orlovsky 1972c; Andersson et al. 1978; Grillner and Rossignol 1978), i.e., it affects the whole synergism of stepping. In contrast, disturbances of the knee and ankle movements are "local". For example, Fig. 54A shows that the delay in the ankle flexion does not influence the hip movement. Thus, according to this hypothesis, the rhythmical activity of RbS neurons is determined by the signals coming to the cerebellum from the central spinal mechanism of stepping and not by the signals on the activity of the executive motor apparatus of the limb.

Direct data indicating that the signals coming from the central spinal mechanisms of locomotion participate in generating rhythmical cerebellar output signals were obtained by French physiologists. Viala and colleagues (Viala et al. 1970) have shown that during fictitious locomotion of curarized rabbits (i.e., in the absence of any rhythmical afferent inflow from the limb receptors), neurons of the cerebellar cortex (presumably Purkinje cells) exhibit the rhythmical activity related with "locomotor" activity in the spinal cord. Perret (1973, 1976) found rhythmical activity in the RS and RbS neurons during fictitious locomotion of curarized decorticated cats, the activity being also related to the "locomotor" discharges in muscle nerves.

These data suggest that signals from the central spinal mechanisms of locomotion coming to the cerebellum play an important (if not a crucial) role in generating its output signals.


## 2. Role of Signals Concerning Peripheral and Central Processes During Scratching

A detailed study of the role of signals concerning the central and peripheral processes in generating cerebellar output signals was carried out for the scratch reflex. In Chap. I we emphasized that the activity of spinal mechanisms during fictitious scratching is almost the same as during actual scratching. But in the former case there is no rhythmical afferent inflow from limb receptors to the spinal cord. Thus, comparing the behaviour of neurons of descending tracts during actual scratching, described in Chap. III, with their behaviour during fictitious scratching, described in the present section, one can estimate the contribution of signals on peripheral and central processes to rhythmical modulation of these neurons.

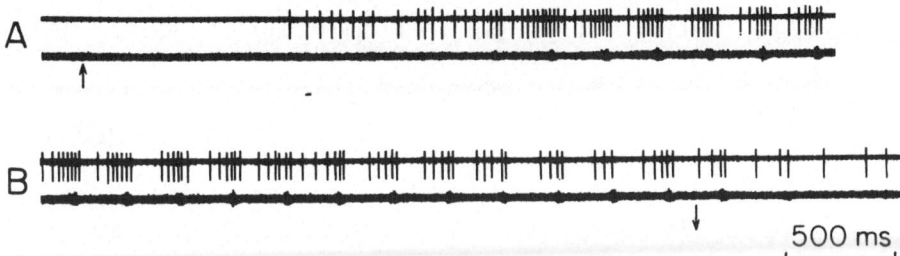

**Fig. 55 A, B.** Activity of a VS neuron during fictitious scratching (**B** is the continuation of **A**). The *arrows* indicate switching on and off of the C2 stimulation. The *lower trace* is the gastrocnemius ENG (Arshavsky et al. 1978c)

During fictitious scratching, most of the VS neurons exhibit the rhythmical activity related to that of motoneurons (Arshavsky et al. 1975b, 1978c). Figure 55A, B gives a typical example of the activity of a VS neuron. During the latent period of the fictitious scratch reflex the neuron is tonically activated. During scratching it is rhythmically active, the maximum excitation being simultaneous with the beginning of the activity of extensor motoneurons. As in actual scratching, the discharge frequency in the burst in most neurons is 30–100 pulses s$^{-1}$. Figure 56A shows the phase distribution of VS neurons, and Figure 56B the frequency curve for the "average" VS neuron. Comparing these graphs with those for actual scratching (Fig. 41), one notices a striking similarity: in both cases the majority of neurons is active in the second half of the L-phase and in the S-phase. Not only the phases

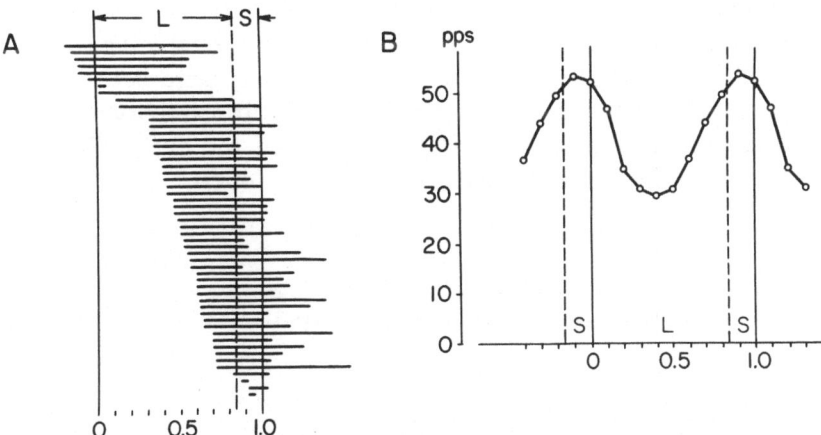

**Fig. 56 A, B.** Relationship between the activity of VS neurons and the phase of the scratch cycle (fictitious scratching). **A** The phase distribution of the 53 VS neurons. **B** The discharge frequency of the "average" VS neuron as a function of the phase of the cycle (Arshavsky et al. 1978c)

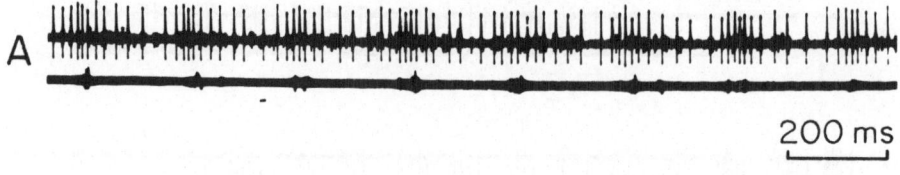

**Fig. 57A, B.** Activity of a VS neuron recorded initially during actual scratching (**A**) and then during fictitious scratching (**B**) after Flaxedil injection (5 mg kg$^{-1}$, i.v.). In **A**, the *lower trace* is the gastrocnemius EMG, and in **B**, the ENG. Amplification in **B** was changed as compared to **A** (Arshavsky et al. 1978c)

of activity are distributed in a similar way, but also the values of discharge frequencies are (on the average) almost the same, as can be seen from the frequency curves. Finally, in both cases the neurons are tonically activated during the latent period of rhythmical scratching (cf. Figs. 55A and 39F, G).

Figure 57 shows a neuron recorded during actual scratching (A) and then during fictitious scratching, after the animal had been immobilized (B). Since the firing pattern persisted after immobilization, one can conclude that not only the behaviour of the population of VS neurons, but also that of individual units is not affected by the elimination of the rhythmical afferent inflow from moving limbs. The rhythmical activity of VS neurons during fictitious scratching, like during actual scratching, completely depends on the cerebellum: in decerebellate animals it is absent.

Most RS neurons were not active during fictitious scratching, as they were during actual scratching. Only a few neurons were active, their discharge was modulated in the rhythm of motoneuron activity (Pavlova 1977). These units fired mainly in the L-phase of the cycle (Fig. 58) like the ones recorded during actual scratching (Fig. 43G).

**Fig. 58A, B.** Activity of two RS neurons during fictitious scratching. The *lower trace* is the gastrocnemius ENG (Pavlova 1977).

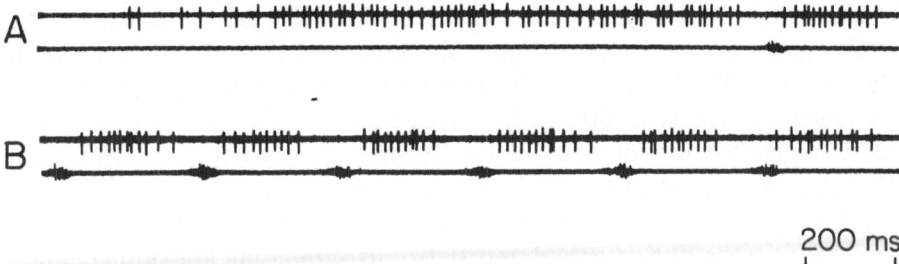

200 ms

**Fig. 59A, B.** Activity of a RbS neuron during fictitious scratching (**B** is the continuation of **A**, the pinna stimulation was started at the beginning of **A**). The *lower trace* is the gastrocnemius ENG (Arshavsky et al. 1978e)

In RbS neurons, the tonic activity usually increases during the latent period of the fictitious scratch reflex. By the beginning of rhythmical generation the firing rate was, on average, 3.5 times higher than at rest. During scratching, the rhythmical activity of RbS neurons develope (Arshavsky et al. 1978e). They periodically generate bursts of impulses (Fig. 59). The firing rate in the bursts usually ranges from 30 to 100 pulses $s^{-1}$, the same as during actual scratching.

Figure 60A shows the phase distribution of RbS neurons during fictitious scratching. It is similar to that during actual scratching (Fig. 48A): in both cases the bursts are widely distributed throughout the

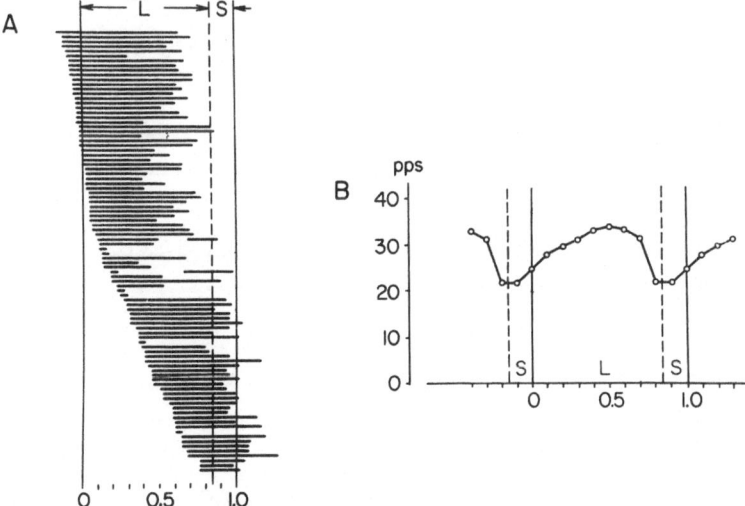

**Fig. 60A, B.** Relationship between the activity of RbS neurons and the phase of the scratch cycle (fictitious scratching). **A** The phase distribution of 94 RbS neurons. **B** The discharge frequency of the "average" RbS neuron as a function of the phase of the cycle (Arshavsky et al. 1978e)

cycle. However, though the neurons begin firing in all phases of the
cycle, most of them stop firing in the second half of the L-phase or in
the S-phase. As a result, as one can see from the frequency curve for the
"average" neuron (Fig. 60B), the overall activity of the whole population
of RbS neurons is somewhat higher in the L-phase than in the S-phase, a
fact also observed during actual scratching (Fig. 48B).

    Though average characteristics of the behaviour of the population
of RbS neurons during actual and fictitious scratching are similar, some

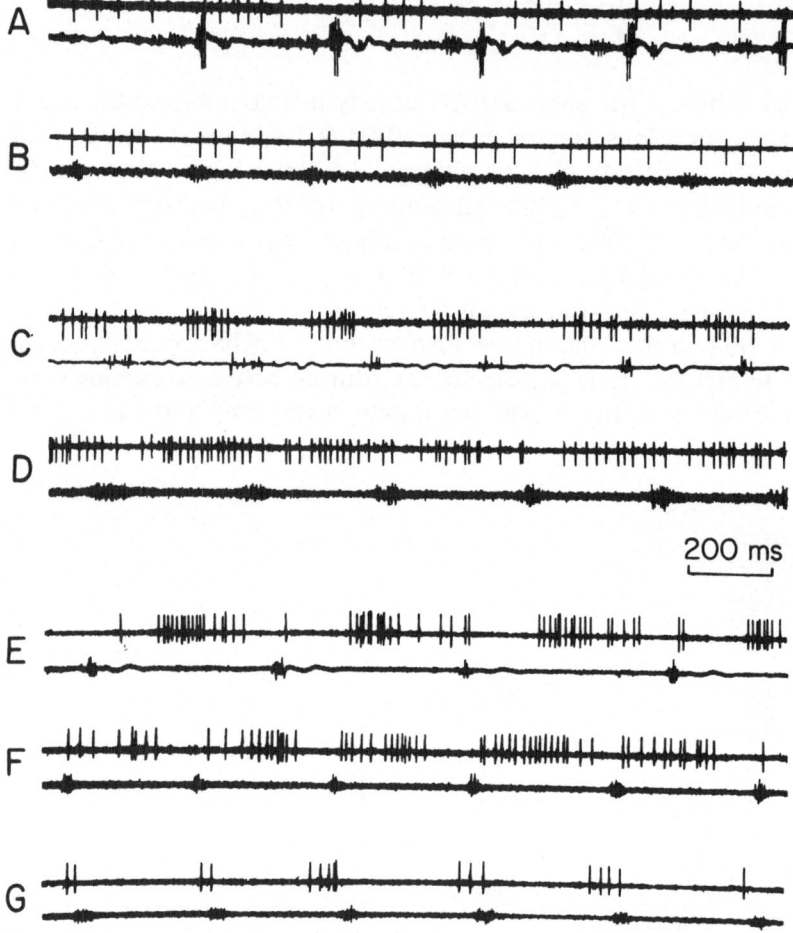

**Fig. 61 A–G.** Activity of three RbS neurons (**A, B; C, D** and **E–G**, respectively)
recorded initially during actual scratching (**A, C** and **E**) and then during fictitious
scratching (**B, D, F** and **G**) after Flaxedil injection (5 mg kg$^{-1}$, i.v.). The *lower
trace* in **A, C** and **E** is the EMG of contralateral m.gastrocnemius; in **B, D** and **F** it
is the ENG of the corresponding nerve. In **G**, the fictitious scratch reflex was evoked
on the ipsilateral (to a neuron) side, and the ENG of ipsilateral n.gastrocnemius is
shown (Arshavsky et al. 1978e)

differences can be found on comparing the activity of individual cells before and after immobilization of the animal. Figure 61 shows three RbS neurons tested first during actual and then during fictitious scratching. It can be seen that in both cases, all three neurons are most active in the L-phase. However, the burst of the first neuron is somewhat shifted from the beginning of the cycle towards the end of the previous one (cf. A and B); the switching on of the second neuron is shifted from the middle of the cycle towards its beginning which makes the duration of the burst about twice as long (cf. C and D); and the burst of the third neuron is considerably shifted towards the beginning of the cycle (cf. E and F).

The rhythmical activity of RbS neurons during fictitious scratching is determined by signals coming to them from the cerebellum since in decerebellate animals only tonic activation of these neurons was observed.

Thus, the pattern of rhythmical activity of the neurons of descending tracts during fictitious scratching is, to a first approximation, similar to that during actual scratching. One can, therefore, conclude that the main role in generating the cerebellar output signals which provide rhythmical activity of the neurons of descending tracts is played by signals from intraspinal processes and not from the activity of the executive motor apparatus.

## 3. Role of Signals Coming via VSCT and SRCP

The information on intraspinal processes reaches the cerebellum via two pathways, the VSCT and the SRCP (see Chap. II). To estimate the contribution from each of these pathways to generating the cerebellar output signals, the effects of their separate transections upon the rhythmical activity of descending tracts were studied during fictitious scratching. Such experiments could be performed since the VSCT and the SRCP ascend along different sides of the spinal cord. The VSCT fibres cross the midline at the lumbar level, ascend in the contralateral lateral funiculus and then again cross the midline as they enter the cerebellum. In contrast, spino-reticular fibres, terminating in the LRN, ascend in the ipsilateral lateral funiculus of the spinal cord. As it has been already mentioned, the central pathway of the scratch reflex crosses the midline twice: most of its fibres pass to the opposite side at the upper cervical level and then return to the initial side at the thoracic level (Deliagina 1977). For this reason, to interrupt the VSCT, a contralateral hemi-section of the spinal cord was performed at the L1 level. To interrupt the SRCP, transection of the ipsilateral lateral funiculus was performed at the C4–C5 level. It has been shown above that such a transection of the lateral funiculus eliminates the rhythmical activity of LRN neurons (Fig. 34A, B). Effects of the VSCT and SRCP transections were studied only for VS and RbS neurons since most RS neurons were not active during scratching.

Figure 62A, B shows the activity of a VS neuron before (A) and after (B) the contralateral hemisection of the spinal cord. One can see that rhythmical discharge modulation almost completely disappears after hemisection. Similar results were obtained for RbS neurons (Fig. 62C, D).

Transection of the ipsilateral lateral funiculus in which the spinoreticular fibres ascend has much weaker influence upon the rhythmical activity of VS and RbS neurons (Fig. 63).

Thus, of the two pathways transmitting messages on the activity of the central spinal mechanisms to the cerebellum, the VSCT is of major importance for rhythmical modulating of VS and RbS neurons. In contrast, the SRCP does not play any important role in the control of neurons of these two descending tracts.

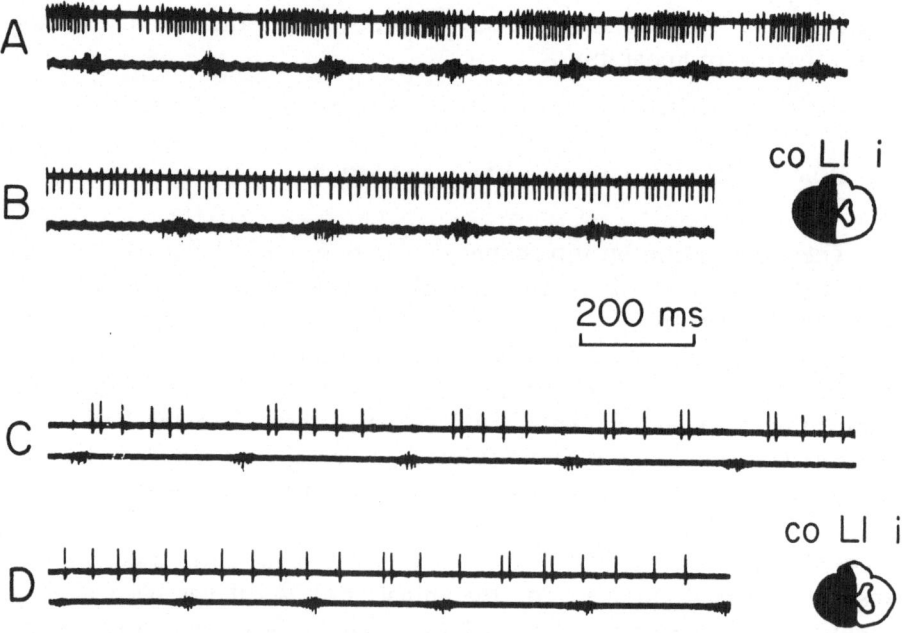

**Fig. 62 A–D.** Effects of contralateral hemisection of the spinal cord (at the *L1* level) upon the activity of VS (**A, B**) and RbS (**C, D**) neurons during fictitious scratching. Both neurons were recorded before (**A, C**) and after (**B, D**) hemisection. The *lower trace* is the gastrocnemius ENG. The extent of lesions is shown as *black areas* on the spinal cord cross-sections (*i* and *co*: ipsi- and contralateral sides in relation to scratching) (Arshavsky et al. 1978c, e)

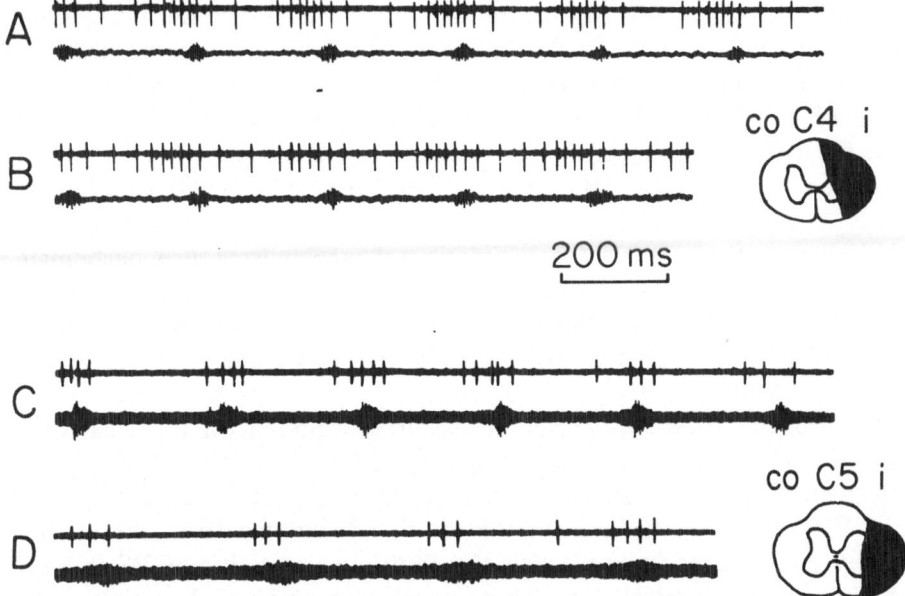

**Fig. 63 A–D.** Effects of transection of the ipsilateral lateral funiculus of the spinal cord (at the *C4–C5* level) upon the activity of VS (**A, B**) and RbS (**C, D**) neurons during fictitious scratching. Both neurons were recorded before (**A, C**) and after (**B, D**) transection. The *lower trace* is the gastrocnemius ENG (Arshavsky et al. 1978c, e)

In Chap. II it was demonstrated that during scratching, rhythmical activity appears not only in the VSCT projecting to the ipsilateral cerebellum, but also in the VSCT projecting contralaterally, although in the latter case, the rhythmical activity of VSCT neurons is considerably weaker (Fig. 26). While recording VS and RbS neurons, it was also found that a weak rhythmical activity appears during fictitious scratching evoked on the opposite (in relation to the VS and RbS tracts) side (Fig. 61G). This activity may be caused, to some extent, by signals reaching the contralateral part of the cerebellum via the corresponding VSCT.

## 4. Conclusion

The study of the activity of neurons of descending tracts during fictitious scratching has shown that signals coming to the cerebellum from the central spinal mechanisms play the decisive role in controlling the activity of these neurons. During fictitious scratching, i.e., in the absence of any rhythmical afferent signals from limb receptors, the rhythmical activity of VS, RS and RbS neurons is as pronounced as during actual scratching. Thus, the signals coming from the rezeptors of the moving limbs are of secondary importance for generating the rhythmical activity of neurons of descending tracts.

According to the data obtained by Perret (1973, 1976), RS and RbS neurons are rhythmically active during fictitious locomotion. This means that signals from the central mechanisms also participate in producing the cerebellar output signals during locomotion. The experiments with the "local" disturbances of stepping movements (Orlovsky 1972c) have demonstrated that considerable changes of the afferent flow from the limb do not affect the rhythmical activity of RbS neurons. This result can be easily interpreted on the basis of a hypothesis on the crucial role of signals coming from the central spinal mechanisms in producing the rhythmical activity of RbS neurons.

The results on the crucial role of signals coming from the central spinal mechanisms in producing the rhythmical activity of VS and RS neurons fit anatomical data. These neurons receive signals mainly from the medial part of the cerebellum (the vermis and fastigial nuclei) in which the VSCT and SRCP, transmitting signals on intraspinal processes, terminate. On the other hand, terminals of the DSCT, which convey messages concerning the activity of the executive motor apparatus, are rather sparse in this part of the cerebellum. No wonder that information concerning intraspinal processes plays the leading role in modulating VS and RS neurons.

Unlike Deiters' nucleus and the medial reticular formation, the red nucleus is supplied by fibres originating from the interpositus nucleus which, in turn, is under the control of the intermediate zone of the cerebellar cortex. Apart from the VSCT and SRCP, most of the DSCT fibres terminate in this zone. Thus, one might suggest that signals on the activity of the executive motor apparatus play an important role in the control of RbS neurons. However, the experiments did not confirm this suggestion. They demonstrated that the rhythmical activity of RbS neurons is as pronounced during fictitious scratching as during actual scratching; immobilization of the cats only slightly affects the phase of the activity of neurons in the cycle. Thus, the rhythmical activity of RbS neurons, like that of VS and RS neurons, is almost completely determined by signals coming from central spinal mechanisms.

Signals on the activity of central mechanisms are important for generating the cerebellar output signals not only in those movements in which the spinal level of the nervous system is the leading one, but also when the higher levels are leading. This was demonstrated in chronic experiments on monkeys trained to perform precise movements (flexion and extension of the arm at the elbow joint (Bioulac and Lamarre 1977). Such movements are accompanied by changes in the activity of Purkinje cells preceding the beginning of movements. After limb deafferentation the Purkinje cells continued to reveal changes in activity in relation to the movements. Under similar experimental conditions it was found that changes of the activity of Purkinje cells and of interpositus and dentate neurons, related with arm movements, are not affected by "local" disturbances of these movements (Harvey et al. 1977, 1979). This finding suggests that changes of the activity of cerebellar neurons are of a central origin. Ghez and Vicario (1978) studied discharges of red nucleus neurons during motor activity in freely moving cats. They concluded that the activity of these neurons, related to movements, is caused by signals from motor centres, but not from the receptors of the executive motor apparatus. It should be noted, however, that as distinct from experiments on decerebrate animals, the experiments on animals with the intact nervous system do not permit any conclusions concerning the concrete sources of central signals affecting cerebellar output neurons. The signals might not only come from the spinal cord, but also from higher motor centres, for instance, from the cerebral cortex.

However, a different point of view also exists concerning the sources of modulation of neurons of the red and interpositus nuclei, based on the following data. In chronic experiments on cats it was found that changes of the activity of IN neurons related to forelimb movements are strongly reduced after deafferentation of the forelimb (Soechting et al. 1978). In chronic experiments on monkeys Otero (1976) found that during arm movements the discharge frequency in the majority of the red nucleus neurons changed only after the beginning of muscle activity. These data were interpreted as proof of the existence of powerful afferent influences upon the neurons of the interpositus and red nuclei. So far there is no explanation for the contradiction between the data concerning the role of peripheral and central factors in generating cerebellar output signals.

Of the two pathways (VSCT and SRCP) transmitting messages from the central spinal mechanisms, the VSCT is of major importance for producing the rhythmical activity of VS and RbS neurons. The different contribution of these two pathways to generating cerebellar output signals will be considered in Chap. V.

# V Activity of Cerebellar Neurons

As we saw in the previous chapters, during locomotion and scratching spino-cerebellar pathways convey rhythmical signals which provide the basis for the generation of the cerebellar output signals. In turn, the cerebellar output signals perform rhythmical modulation of the activity of neurons of descending tracts. In the present chapter the activity of Purkinje cells and of neurons of the cerebellar nuclei will be described. We shall also continue the analysis of the role of different input signals in generating cerebellar output signals. Then, the activity of cerebellar neurons will be compared with the activity of neurons of ascending and descending tracts.

## 1. Purkinje Cells

Purkinje cells are the output neurons of the cerebellar cortex. These large cells have wide branching dendrites. The cell bodies form the ganglionic layer located between the molecular and granular layers of the cortex. The Purkinje cells receive excitatory inputs from the granule cells and climbing fibres, and inhibitory inputs from neurons of the molecular layer (Fig. 13B). Axons of the Purkinje cells form inhibitory synapses on neurons of the cerebellar nuclei and of Deiters' nucleus (see Chap. III).

Purkinje cells have high background activity (Dow and Moruzzi 1958; Eccles et al. 1967; Thach 1967, 1968, 1970b; Arshavsky et al. 1971b, 1984b; Orlovsky 1972d). This activity is very irregular: the discharge frequency can spontaneously increase to 100–200 pulses $s^{-1}$ or decrease to zero (Fig. 65A, C). Due to the background activity, Purkinje cells exert a tonical inhibitory influence upon the neurons of the cerebellar nuclei and of Deiters' nucleus. Inhibition of Purkinje cells or cooling of the cerebellar cortex causes disinhibition of neurons of Deiters' and cerebellar nuclei (Eccles et al. 1967; Ito et al. 1968b; Rosén and Scheid 1972).

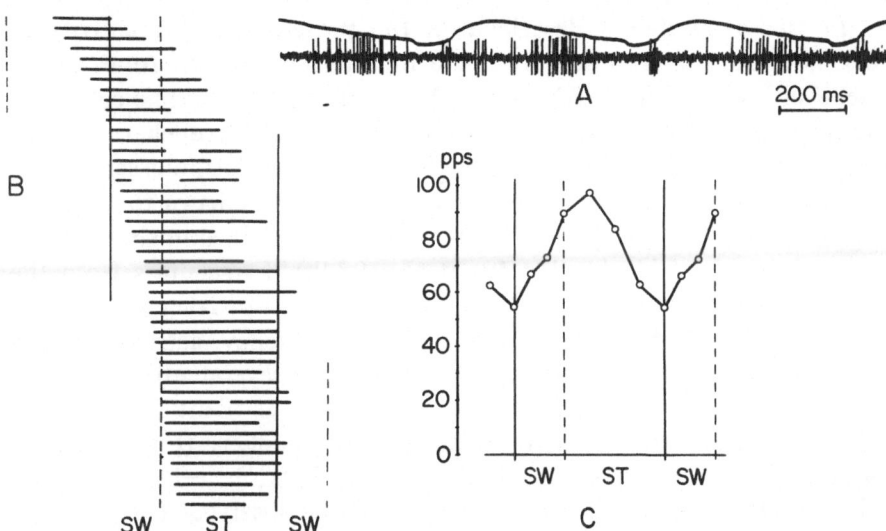

**Fig. 64 A–C.** Activity of Purkinje cells during locomotion. **A** Purkinje cell from the pars intermedia of the anterior lobe (point 2 in Fig. 65G), the *upper trace* is the movement of the ipsilateral hindlimb (protraction – up). **B** The phase distribution of 49 Purkinje cells. **C** The discharge frequency of the "average" Purkinje cell as a function of the phase of the step cycle (Orlovsky 1972d)

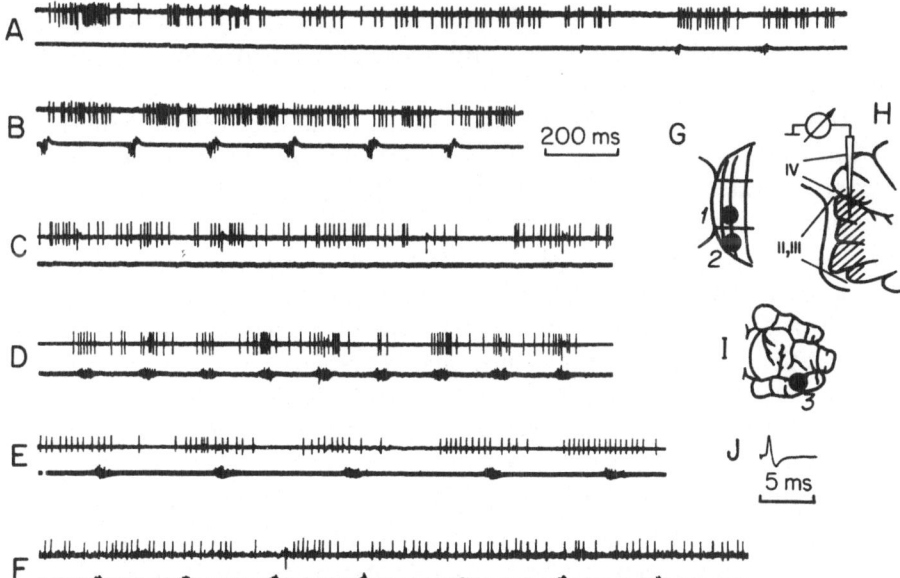

**Fig. 65 A–J.** Activity of Purkinje cells during fictitious scratching. **A, B** A cell from the pars intermedia of the anterior lobe (**B** is the continuation of **A**). **C, D** A cell from the vermis of the anterior lobe (**C**, background activity; **D** scratching). **E** A cell from the vermis (its antidromic response to stimulation of Deiters' nucleus is shown in **J**). **F** A cell from the paramedian lobule. In all the records, the *lower trace* is the gastrocnemius ENG. **G, I** Sites of insertion of the microelectrode in the hindlimb areas of the vermis (*1*), pars intermedia (*2*) and paramedian lobule (*3*). The hindlimb area in the vermis (*lobules II, III*) is *hatched* in **H** (Arshavsky et al. 1984b)

In the experiments described in the present chapter, activity in Purkinje cells was recorded from the hindlimb projection zones of the cerebellar cortex (Fig. 13A). During locomotion, the Purkinje cells were recorded only from the pars intermedia of the cerebellar anterior lobe (point 2 in Fig. 65G) (Orlovsky 1972d). During fictitious scratching, they were recorded from the vermis and pars intermedia of the anterior lobe (points 1 and 2 in Fig. 65G) as well as from the paramedian lobule (point 3 in Fig. 65 I) (Arshavsky et al. 1984b). Large units from the ganglionic layer were considered as Purkinje cells; this layer is easily recognized due to the appearance of high frequency activity with a relatively large amplitude which is generated by several adjacent units. A part of the Purkinje cells from the vermis was identified by antidromic response to stimulation of Deiters' nucleus (Fig. 65J). In about half of the units considered here as Purkinje cells, "complex spikes" (marked by dots in Fig. 70 A–C) could be observed. The complex spikes are generated by Purkinje cells in response to impulses arriving via climbing fibres (see Chap. II). According to the common point of view, these spikes are a reliable criterion for Purkinje cell identification. However, many units considered here as Purkinje cells (including those which responded antidromically to stimulation of Deiters' nucleus) did not generate complex spikes, as the neuron in Fig. 65E. Other authors also recorded Purkinje cells (identified by their antidromic response to stimulation of the cerebellar white matter) that generated no complex spikes (Latham and Paul 1971).

During locomotion, the discharge of most Purkinje cells is modulated in the rhythm of stepping (Fig. 64A). Usually the Purkinje cells fire in bursts separated by periods of silence. The discharge frequency within the bursts usually reaches 100–150, sometimes up to 200–250 pulses $s^{-1}$. Figure 64B shows the phase distribution of Purkinje cells. One can see a great variety of burst positions in the cycle. Nevertheless, the overall activity of the whole population of Purkinje cells has maximum at the beginning of the stance phase (Fig. 64C).

The rhythmical activity of Purkinje cells from the pars intermedia of the cerebellar anterior lobe was also observed during locomotion in freely moving cats (McElligott 1976).

Activity of Purkinje cells during fictitious scratching was examined in more detail. In cells from the vermis and pars intermedia of the cerebellar anterior lobe, the background activity usually does not change during the latent period of the fictitious scratch reflex (which is evoked on the ipsilateral side) (Fig. 65A). When the rhythmical bursts of ENG appear, rhythmical modulation of the discharge of Purkinje cells begins as well (Fig. 65B, D, E). As in locomotion, cells generate bursts of impulses separated by periods of silence, the firing rate in the bursts being 50–150 (sometimes up to 200–300) pulses $s^{-1}$. Higher

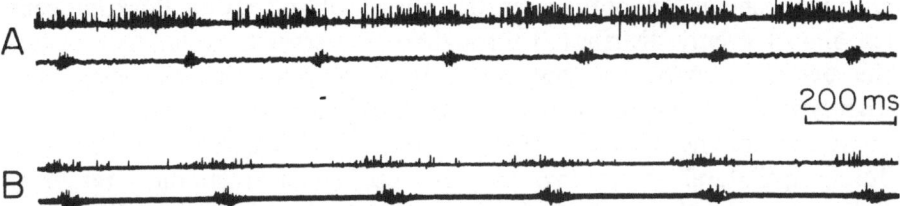

**Fig. 66A, B.** Simultaneous recording of several Purkinje cells during fictitious scratching. Recordings **A** and **B** were performed in different experiments (Arshavsky 1984b)

frequencies were more common for the cells active at the end of the cycle (Figs. 65D and 71A).

As mentioned above, the resting discharge of Purkinje cells is very irregular (Fig. 65A, C). Similar irregulartiy of the rhythmical activity is often observed during locomotion and scratching. Figures 65B and D show that both the burst duration and the discharge rate in the burst, as well as the inhibition between the bursts can vary considerably in successive cycles.

Sometimes the activities of several neighbouring Purkinje cells were recorded simultaneously during fictitious scratching by the same micro-electrode (Figs. 66 and 70B). In these cases Purkinje cells fired more or less in-phase.

Phase distributions of Purkinje cells from the vermis and pars intermedia are shown in Figs. 67A and 68A, respectively. The distributions

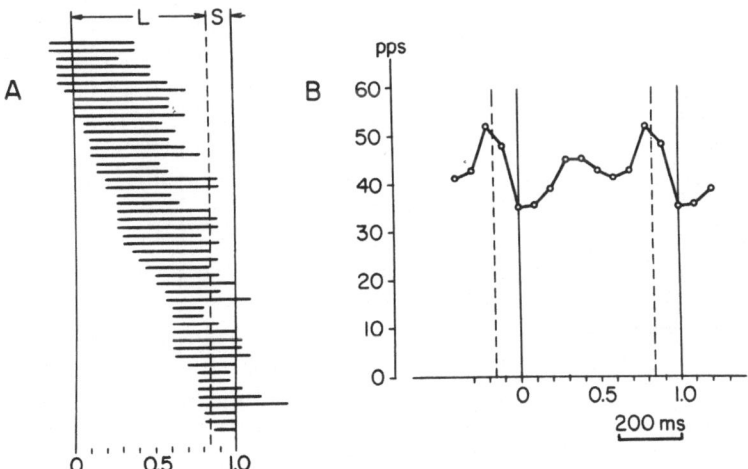

**Fig. 67A, B.** Relationship between the activity of Purkinje cells from the vermis of the anterior lobe and the phase of the cycle during fictitious scratching. **A** The phase distribution of 49 cells. **B** The discharge frequency of the "average" Purkinje cell as a function of the phase of the cycle (Arshavsky et al. 1984b)

are quite similar: in both cases the phases of activity of Purkinje cells are almost evenly distributed throughout the scratch cycle. As a result, the overall activities of the populations of Purkinje cells vary only slightly in the course of the cycle (Figs. 67B and 68B).

A part of the Purkinje cells from the vermis was recorded not only during ipsilateral scratching, but also during contralateral scratching. In the latter case the rhythmical activity was considerably lower (Fig. 69).

In contrast to Purkinje cells from the cerebellar anterior lobe, cells from the paramedian lobule usually exhibited no rhythmical modulation of the discharge during fictitious scratching (Fig. 65F); only in a few units could weak rhythmical changes of the firing rate be observed.

The rhythmical activity of Purkinje cells during locomotion and scratching described above was determined by rhythmical changes of the rate of generation of "simple spikes", i.e., it reflected the rhythmical input through the mossy fibres. In contrast, the generation of "complex spikes" in most Purkinje cells was not linked with the rhythm of locomotion or scratching (Fig. 70B, D). Only in a few cells was a marked correlation between the generation of "complex spikes" and the phase of the cycle observed (the most impressive example is shown in Fig. 70C, E). This result confirms the conclusion made in Chap. II that the SOCP does not participate in the transmission of rhythmical signals during locomotion and scratching.

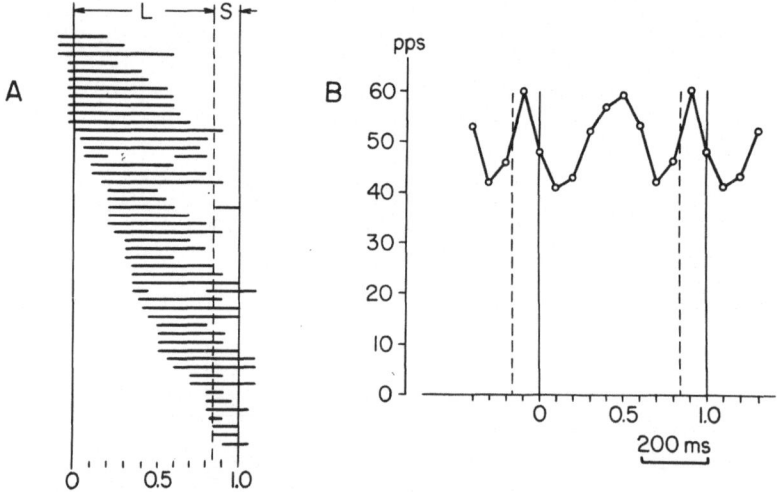

**Fig. 68A, B.** Relationship between the activity of Purkinje cells from the pars intermedia of the anterior lobe and the phase of the cycle during fictitious scratching. **A** Phase distribution of 49 cells. **B** The discharge frequency of the "average" Purkinje cell as a function of the phase of the cycle (Arshavsky et al. 1984b)

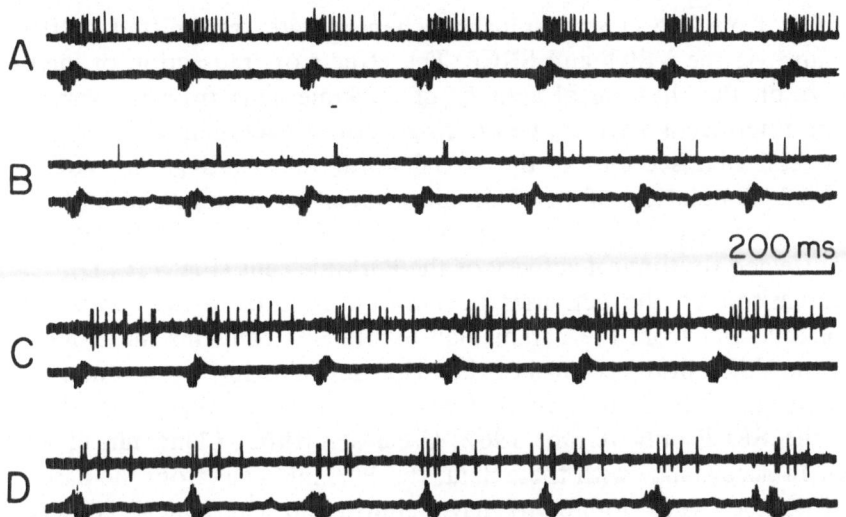

**Fig. 69 A–D.** Activity of two Purkinje cells from the vermis (**A, B** and **C, D**, respectively) during ipsilateral (**A, C**) and contralateral (**B, D**) fictitious scratching. The *lower trace* is the ENG of ipsilateral (**A, C**) and contralateral (**B, D**) n.gastrocnemius (Arshavsky et al. 1984b)

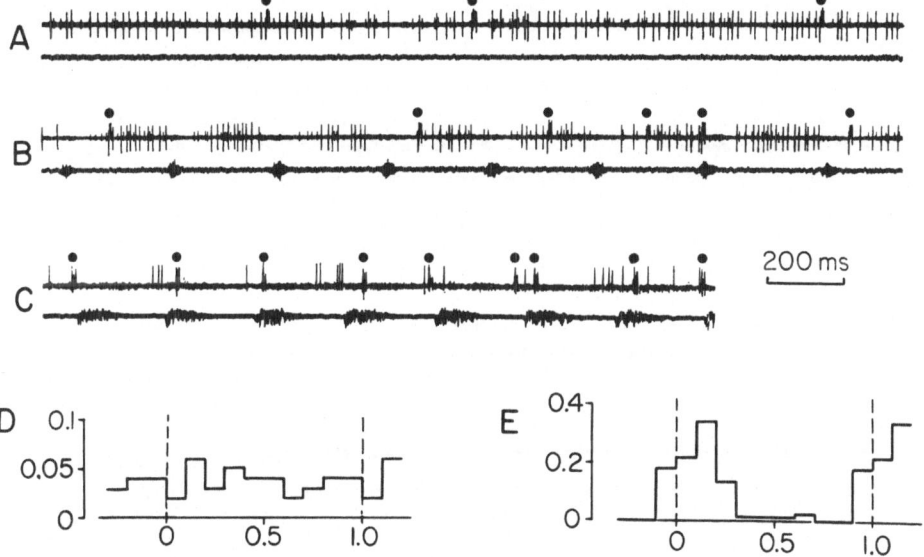

**Fig. 70 A–E.** Generation of "complex spikes" by Purkinje cells. **A, B** The background activity (**A**) and activity during fictitious scratching (**B**) of a Purkinje cell. The *lower trace* is the gastrocnemius ENG. **C** Activity of another Purkinje cell during fictitious scratching. The *lower trace* is the tibialis anterior ENG. The "complex spikes" are marked by *dots*. **D** The "probability" of generating "complex spikes" in various phases of the scratch cycle for the cell recorded in **B** (calculated over 100 successive cycles). **E** The same for the cell shown in **C**. *Ordinate* in **D** and **E** is the mean number of "complex spikes" generated in each 0.1 part of the cycle (Arshavsky et al. 1984b)

Signals from the central spinal mechanisms are transmitted to the cerebellum via the VSCT and SRCP. The effects of transection of these pathways on the rhythmical activity of Purkinje cells from the vermis and pars intermedia were studied during fictitious scratching. After the contralateral hemisection of the spinal cord, interrupting the VSCT, rhythmical modulation of the Purkinje cells either disappeared or considerably decreased (Fig. 71A, B), whereas interruption of the SRCP hardly affected rhythmical activity of the Purkinje cells (Fig. 71C, D).

Comparison of the activities of Purkinje cells from the cerebellar anterior lobe and from the paramedian lobule also confirms the crucial role of the VSCT in the control of their rhythmical firing. In the anterior lobe, both the VSCT and SRCP terminate, while in the paramedian lobule the SRCP only (Grant 1962; Oscarsson 1965; Clendenin et al. 1974a). In accordance with these data, the Purkinje cells from the paramedian lobule were found to have almost no rhythmical activity (Fig. 65F).

The data described in the present and preceding chapters have shown that in spite of the fact that the SRCP fibres are widely branching

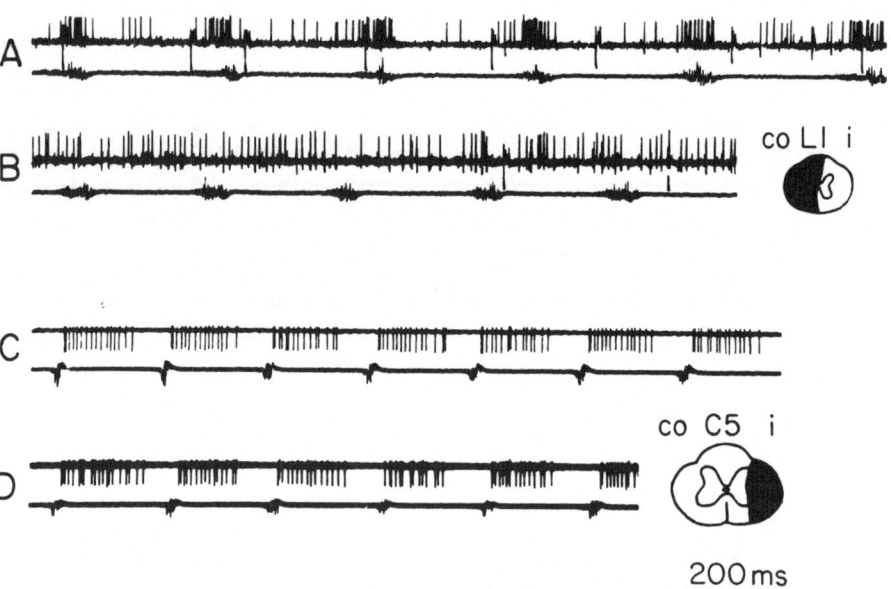

Fig. 71 A—D. Effects of contralateral hemisection of the spinal cord and transection of the ipsilateral lateral funiculus upon the activity of Purkinje cells during fictitious scratching. A, B A cell was recorded before (A) and after (B) contralateral hemisection at the *L1* level. C, D A cell was recorded before (C) and after (D) transection of the ipsilateral lateral funiculus (at the *C5* level). The *lower trace* is the gastrocnemius ENG. The extent of lesions is shown to the *right* of B and D (Arshavsky et al. 1984b)

in the cerebellar cortex, the signals they transmit have only minor importance for producing the rhythmical activity of Purkinje cells. These data are in accordance with the hypothesis on the different functional meaning of signals coming to the cerebellar cortex through the direct spino-cerebellar tracts and through the SRCP (Arshavsky et al. 1969a, b, 1970, 1971a, 1972c). The hypothesis (based on the study of cortical responses to peripheral stimuli) claims that direct spino-cerebellar tracts (including the VSCT), on the one hand, and the SRCP, on the other, terminate on different groups of granule cells. The VSCT terminates on the granule cells of type I which directly affect the Purkinje cells and thus produce the output signals of the cerebellar cortex. Type II granule cells, on which the SRCP terminates, do not directly affect Purkinje cells but, through Golgi cells, control the efficiency of signals coming via direct tracts. Thus, signals coming via SRCP can exert only indirect influences upon the output signals of the cerebellar cortex.

The hypothesis on two functionally different cerebellar inputs is indirectly supported by the findings described in the present and preceding chapters: signals coming through the VSCT are much more effective than those coming through the SRCP.

## 2. Fastigial Nucleus

### a) General Characteristics

The fastigial nuclei (FN) are the most medially located cerebellar In cats, the FN has a volume of about 12.3 mm³ and contains, on average, 7300 neurons (Palkovits et al. 1977). The FN neurons are of different size. The large and some middle-sized FN neurons are output units, while the rest of the middle-sized and small cells are the interneurons (Snider 1940; Flood and Jansen 1961; Eager 1968; Matsushita and Iwahori 1971a; Fanardjian 1975; Palkovits et al. 1977). Large neurons previal in the rostral part of the nucleus and the small ones in its caudal part. FN neurons fire at rest with frequencies of 10–40 pulses s$^{-1}$ (Arduini and Pompeiano 1957; Eccles et al. 1967, 1974e; Orlovsky 1972e; Antziferova et al. 1980). The background activity of cerebellar nuclei neurons is assumed to be supported by synaptic inflow coming via the collaterals of mossy and climbing fibres (Eccles et al. 1967; Eccles 1973).

FN neurons receive signals mainly from the Purkinje cells located in the vermis, mostly in its medial part (Jansen and Brodal 1940, 1954; Voogd 1964; Eccles et al. 1967; Armstrong 1978). Axons of the Purkinje cells project mainly onto the ipsilateral FN and, to a lesser extent, to the contralateral one (Jansen and Brodal 1954; Eager 1963, 1966; Walberg and Jansen 1964; Fanardjian 1975; Courville and Diakiw 1976). Cortico-nuclear projections in the cerebellum are characterized by enormous divergence and convergence. One Purkinje cell may innervate up to 50 (on the average, 13.5) neurons within a given cerebellar nucleus; as much as 860 Purkinje cells may converge on a single cell of the nucleus (Palkovits et al. 1977).

Axonal terminations of the Purkinje cells form about 62% of the total amount of synapses located on FN neurons (Palkovits et al. 1977). The remaining synapses are formed mainly by collaterals of the afferent fibres coming to the cerebellar cortex. The spinal input to the FN is formed mainly by the collaterals of the VSCT, SRCP and SOCP fibres (Matsushita and Ikeda 1970a, b, 1976; Dietrichs and Walberg 1979; Amatuni 1981). Synapses on the FN neurons are also formed by recurrent collaterals of the fastigiofugal fibres as well as by the interneurons situated in the nucleus.

Stimulation of the nerves or cutaneous receptors of fore- and hindlimbs causes diphasic (excitation followed by inhibition) reactions of FN neurons (Eccles et al. 1974e, f). Analysis of these reactions has revealed that excitation is mediated by the collaterals of the SRCP and SOCP fibres, while the inhibition is relayed by the axons of Purkinje cells. The collaterals of direct spino-cerebellar fibres (the VSCT in particular) do not participate in producing the excitation in FN neurons.

The units responding mainly to stimulation of the forelimb receptors are located more caudally in the FN than those which preferably respond to stimulation of the hindlimb receptors (Eccles et al. 1974b). However, somatotopic discrimination in the FN is not well represented since most of the neurons respond to stimulation of receptors of both the fore- and hindlimbs. Nor do the FN neurons discriminate inputs from symmetrical limbs (Eccles et al. 1974b).

Fibres originating from the FN terminate mainly in the caudal part of the brain stem, in particular, in the Deiters nucleus and in the medial ponto-medullary reticular formation. The fibres originating from the rostral part of the FN project onto the ipsilateral Deiters nucleus and those from the caudal part of the FN onto the contralateral one. These fastigiofugal fibres terminate mainly on small VS neurons (Brodal et al. 1962). Fibres terminating in the medial reticular formation originate from the rostral and, to a smaller extent, from the caudal parts of the FN. These fibres project to the contralateral medial reticular formation and, to a smaller extent, to the ipsilateral one (Walberg et al. 1962; Eccles et al. 1975a; Batton et al. 1977). A few of the FN neurons send axons directly to the spinal cord, but they do not descend farther than the upper cervical segments (Batton et al. 1977; Wilson et al. 1977, 1978). Finally, some of the fibres originating in the caudal part of the FN are directed rostrally and terminate in the tectum and in some thalamic nuclei (Angaut and Bowsher 1970; Batton et al. 1977).

b) Activity of FN Neurons During Locomotion
and Fictitious Scratching

Recording of the activity of FN neurons during locomotion (Orlovsky 1972e) and fictitious scratching (Antziferova et al. 1979b, 1980) was performed from the rostral (magnocellular) part of the FN to which the Purkinje cells from the vermis of the cerebellar anterior lobe are projecting (Fig. 72A, B). To select the units related to the activity of the spinal hindlimb centre, reactions of neurons to passive movements of different limbs were tested, and for further recordings only the units with well pronounced reactions to the hindlimb movements (Fig. 72D) were used. Usually these neurons responded, to some extent, to movements of other limbs as well, which corresponds to the results obtained by Eccles et al. (1974b) on poor somatotopic discrimination in the FN.

During locomotion, the rhythmical activity related to limb movements was observed in most neurons. In the example in Fig. 72 F a neuron fires (about 100 pulses $s^{-1}$) in the swing phase of the step which was typical for most FN neurons. Figure 73A shows their phase distribution, and Fig. 73B the frequency curve for the "average" FN neuron. One can see that the population of FN neurons is maximally active in the swing phase.

The behaviour of the FN neurons was also studied during the fictitious scratch reflex that was evoked on both the ipsi- and contralateral side (in relation to the neuron). With stimulation of the ipsilateral pinna, during the latent period of rhythmical generation, the resting discharge of FN neurons is either unaffected or inhibited (Fig. 74B), When the rhythmical process in the spinal cord arises, the corresponding rhythmical activity appears in most of the neurons (Fig. 74B); the firing rate within the bursts being 30—80, sometimes up to 150 pulses $s^{-1}$. With enhancing of the rhythmical process in the spinal cord, the rhythmical activity of the neurons also increases (cf. A and B in Fig. 74).

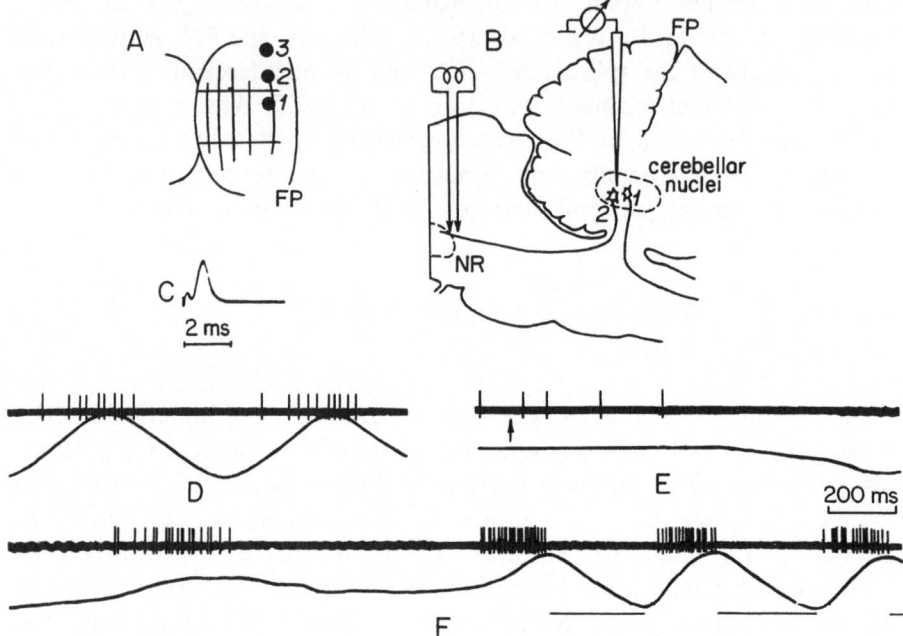

**Fig. 72A, B.** Recording of neurons of cerebellar nuclei and identification of IN neurons. Neurons of the FN were recorded from the rostral part of the nucleus (position of a neuron and site of the microelectrode insertion are marked by number *1*). Neurons of the IN were also recorded from the rostral part of the nucleus and identified by the antidromic response to stimulation of the contralateral red nucleus (*NR*) (position of a neuron and site of the microelectrode insertion are marked by number *2*). Site of insertion of the microelectrode for recording of LN neurons is marked by number *3. FP* − fissura prima. **C** The antidromic response of a IN neuron to stimulation of the red nucleus. **D−F** Responses of a FN neuron to passive movements of the ipsilateral hindlimb (**D**) and its activity during locomotion (**E, F**). **F** is the continuation of **E**. The *arrow* indicates the beginning of stimulation of the locomotor region. The *lower trace* is the movement of the ipsilateral hindlimb (protraction − up); stance phases of the limb are marked by *horizontal lines* in **F** (Orlovsky 1972e)

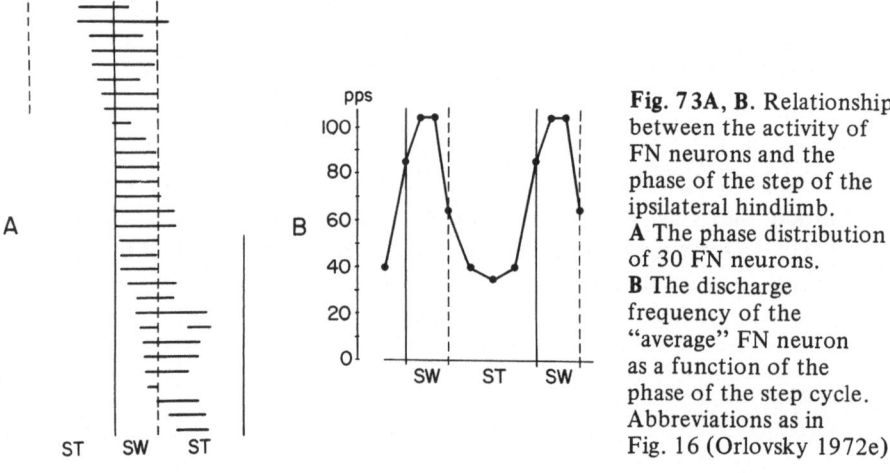

**Fig. 73A, B.** Relationship between the activity of FN neurons and the phase of the step of the ipsilateral hindlimb. **A** The phase distribution of 30 FN neurons. **B** The discharge frequency of the "average" FN neuron as a function of the phase of the step cycle. Abbreviations as in Fig. 16 (Orlovsky 1972e)

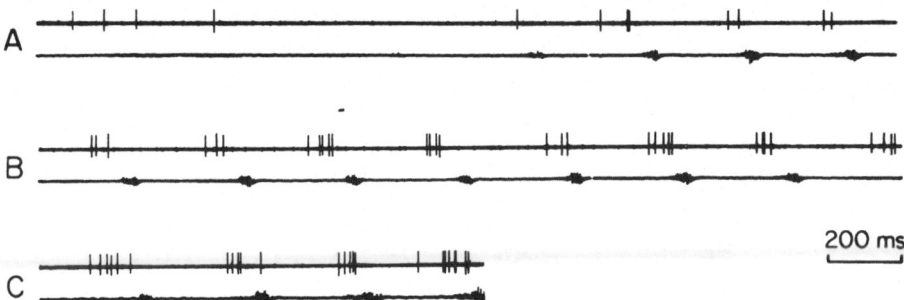

**Fig. 74 A–C.** Activity of a FN neuron during fictitious scratching. **A, B** Ipsilateral scratching (**B** is the continuation of **A**). **C** Contralateral scratching. The *lower trace* is the ENG of ipsilateral (**A, B**) and contralateral (**C**) n.gastrocnemius (Antziferova et al. 1980)

Figure 75A presents the phase distribution of the FN neurons during ipsilateral fictitious scratching. Most neurons fire at the end of the L-phase and in the S-phase; thus, the overall activity of the population of FN neurons is maximum in the S-phase (Fig. 75B) when the extensor motoneurons are active. Thus, the behaviour of FN neurons in this case differs from that in locomotion where they fire mainly in the swing (flexor) phase (Fig. 73B). The reason for this difference is unclear.

The behaviour of FN neurons during contralateral fictitious scratching is similar to that during ipsilateral scratching. In most neurons, neither the burst position in the cycle nor the firing rate in the bursts depend on the side on which the reflex is evoked (cf. Fig. 74B and C). Figure 76 shows the phase distribution (A) and the frequency curve for the "average" FN neuron (B) obtained for contralateral scratching. These graphs are very similar to those for ipsilateral scratching (Fig. 75). One may conclude that when the spinal centre of one of the hindlimbs is rhythmically active, the corresponding rhythmical activity appears in both fastigial nuclei.

The rhythmical activity of FN neurons during fictitious scratching is determined by signals coming to the cerebellum from the central spinal mechanisms through the VSCT and SRCP. To estimate the role of each of these pathways, the effects of their separate transections upon the activity of FN neurons during ipsilateral scratching were studied (Fig. 77). Transection of the ipsilateral lateral funiculus, where the SRCP fibres ascend, results in a considerable reduction of the bursts generated by a neuron (cf. Fig. 77A and B). In contrast, the contralateral hemisection of the spinal cord, resulting in the interruption of the VSCT, leads only to a decrease of the inhibition between the bursts, while the maximum firing rate within the bursts persists (cf. Fig. 77C and D).

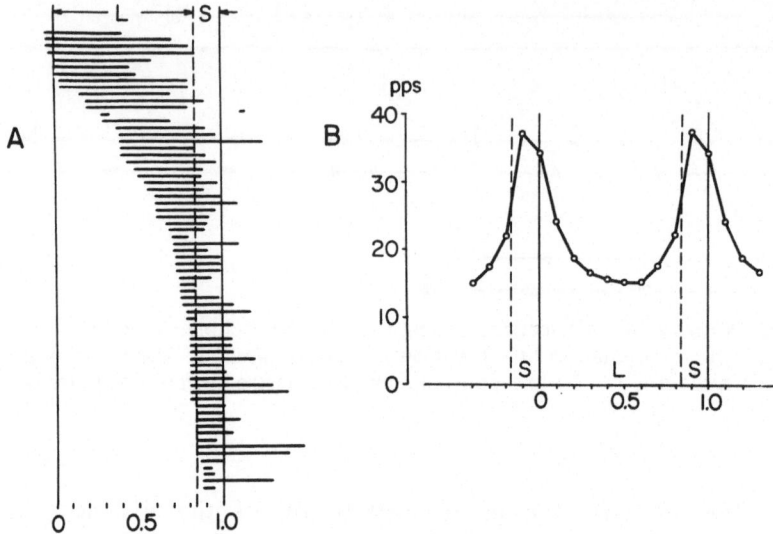

**Fig. 75A, B.** Relationship between the activity of FN neurons and the phase of the cycle during ipsilateral fictitious scratching. **A** The phase distribution of 68 FN neurons. **B** The discharge frequency of the "average" FN neuron as a function of the phase of the cycle (Antziferova et al. 1980)

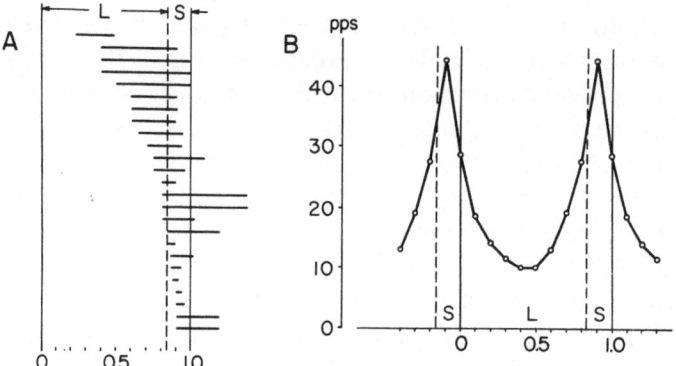

**Fig. 76A, B.** Relationship between the activity of FN neurons and the phase of the cycle during contralateral fictitious scratching. **A** The phase distribution of 25 FN neurons. **B** The discharge frequency of the "average" FN neuron as a function of the phase of the cycle (Antziferova et al. 1980)

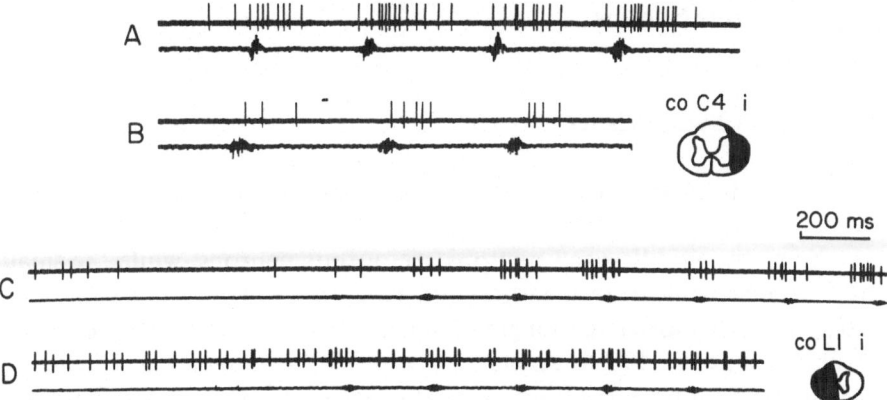

**Fig. 77 A—D.** Effects of transection of the ipsilateral lateral funiculus and of contra-lateral hemisection of the spinal cord upon the activity of FN neurons during fictitious scratching. **A, B** A neuron was recorded before (**A**) and after (**B**) transection of the ipsilateral lateral funiculus at the *C4* level. **C, D** A neuron was recorded before (**C**) and after (**D**) contralateral hemisection at the $L_1$ level. The *lower trace* is the gastrocnemius ENG. The extent of lesions is shown on the *right* in **B** and **D** (Antzi-ferova et al. 1980)

In the preceding section it was shown that signals conveyed by the SRCP practically do not affect the rhythmical activity of Purkinje cells (Fig. 71C, D). Thus, signals conveyed by the SRCP cannot exert rhyth-mical influences upon the FN neurons through the cerebellar cortex, but through the collaterals of reticulo-cerebellar fibres. This is also proven by the fact that most LRN neurons, like FN ones, are active in the S-phase during both ipsilateral and contralateral scratching (Fig. 32).

Since the VSCT contributes to the inhibition of FN neurons between the bursts, one can suppose that influences of the VSCT upon the FN are mediated by Purkinje cells, and that collaterals of the VSCT fibres terminating in the FN do not produce any marked effect. This corresponds to the conclusion of Eccles et al. (1974f) that signals arriv-ing via collaterals of direct spino-cerebellar pathways evoke no responses in the FN neurons.

## 3. Interpositus Nucleus

### a) General Characteristics

In cats, the interpositus nucleus (IN) has a mean volume of 16.9 mm$^3$ and contains approx. 10,000 neurons Palkovits et al. 1977). The large and sometimes the middle-sized neurons are output neurons, while the small units are interneurons (Matsushita and Iwahori 1971b). The IN neurons display a resting discharge ranging from 2–3 to 90 pulses s$^{-1}$ (Orlovsky 1972e; Eccles et al. 1974c; Arshavsky et al. 1980b; Cody et al. 1981).

The main source of fibres terminating in the IN are the Purkinje cells from the intermediate area of the ipsilateral cerebellar hemisphere (Jansen and Brodal 1940, 1954; Voogd 1964; Eccles et al. 1967). The axons of Purkinje cells form about 59% of the synapses located on IN neurons (Palkovits et al. 1977). Other important sources of the synaptic inflow to IN neurons are the collaterals of mossy and climbing fibres. The IN is supplied by collaterals of the VSCT fibres as well as of the SRCP and SOCP fibres (Matsushita and Ikeda 1970a, b, 1976; Eccles et al. 1974c; MacKay and Murphy 1974; Kitai et al. 1977; McCrea et al. 1977; Dietrichs and Walberg 1979).

Peripheral stimulation evokes complex responses in IN neurons consisting of several phases of excitation and inhibition (Rosen and Scheid 1972; Eccles et al. 1974c; Kawaguchi and Ohno 1974; MacKay and Murphy 1974; Allen et al. 1977). Analysis of these responses has shown that inhibition of IN neurons is produced by the Purkinje cells, and excitation by collaterals of the VSCT, SRCP and SOCP fibres.

As described in Chap. II, afferent connections of the pars intermedia are somatotopically organized. However, somatotopic organization in the IN is much less pronounced. Most of the IN neurons respond to stimulation of nerves and receptors of both fore- and hindlimbs (Eccles et al. 1974a, d). The poor somatotopic discrimination of IN neurons is obviously accounted for by wide divergence and convergence of cortico-nuclear connections in the cerebellum: one Purkinje cell innervates up to 50 nuclear neurons and about 860 Purkinje cells converge on one nuclear neuron (Palkovitz et al. 1977). Furthermore, the collaterals of mossy and climbing fibres are widely branching in the IN (Eccles et al. 1967).

The fibres originating from the IN terminate mainly in the contra-lateral red nucleus producing monosynaptic excitation of RbS neurons (see Chap. III). The part of the IN axons innervating the red nucleus extend further into the ventrolateral nucleus of the thalamus which, in turn, projects to the sensorimotor cortex (Eccles et al. 1967; Tsukahara et al. 1967).

## b)  Activity of IN Neurons During Locomotion and Fictitious Scratching

Recording of the activity of IN neurons during locomotion (Orlovsky 1972e) and fictitious scratching (Arshavsky et al. 1980b) was performed from the rostral part of the IN. Only the units easily responding to passive movements of the ipsilateral hindlimb were tested. Neurons were identified by the antidromic response to stimulation of the contralateral red nucleus (Fig. 72 A–C).

During locomotion, IN neurons behave in almost the same way as FN neurons. Most of them exhibit the rhythmical activity related to stepping movements: they fire in bursts separated by periods of silence, the discharge frequency within the bursts being 50–150 pulses s$^{-1}$.

Figure 78A shows the phase distribution of IN neurons: most of them come into operation in the second half of the stance phase and at the beginning of the swing phase; the neurons are "switched off" in the swing phase or at the beginning of the stance phase. Correspondingly the overall activity of the population of IN neurons is maximum at the beginning of the swing phase and minimum in the stance phase (Fig. 78B).

Let us now consider the behaviour of IN neurons during the fictitious scratch reflex. When the reflex is evoked on the ipsilateral (to the neuron) side, a tonic activation in most neurons is observed during the latent period of the rhythmical generation. Then the rhythmical activity of

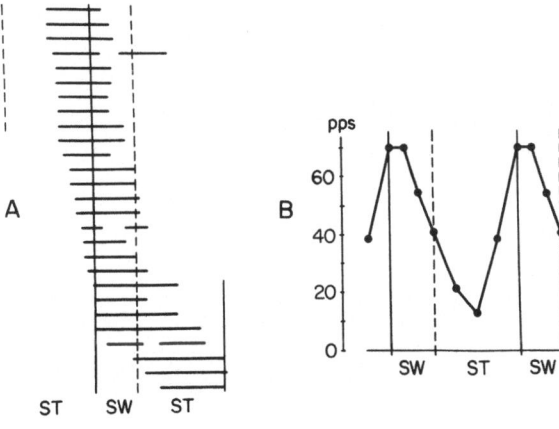

Fig. 78A, B. Relationship between the activity of IN neurons and the phase of the step of the ipsilateral hindlimb. A The phase distribution of 27 IN neurons. B The discharge frequency of the "average" IN neuron as a function of the phase of the step. Abbreviations as in Fig. 16 (Orlovsky 1972e)

neurons appears (Fig. 79A, B, D). They generate bursts of impulses with discharge frequencies in the bursts of 50–105 pulses s$^{-1}$. The phases of activity of neurons in the cycle are diverse (Fig. 80A). Nevertheless, the whole population is maximally active in the L-phase (Fig. 80B). During contralateral scratching IN neurons display only weak rhythmical activity (Fig. 79C) or fire tonically (Fig. 79E).

The rhythmical activity of IN neurons is mainly determined by signals coming through the VSCT. After the contralateral hemisection of the spinal cord (interrupting the VSCT) rhythmical modulation of the discharge disappears (cf. Fig. 81A and B). On the other hand, transection of the ipsilateral lateral funiculus (interrupting the SRCP) hardly affects the

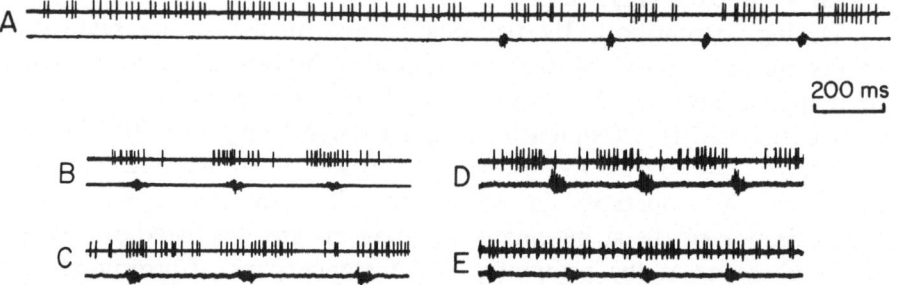

**Fig. 79 A–E.** Activity of three IN neurons (**A**; **B, C** and **D, E** respectively) during fictitious scratching. **B–E** Activity of two neurons during ipsilateral (**B, D**) and contralateral (**C, E**) scratching. The *lower trace* is the ENG of ipsilateral (**A, B, D**) and contralateral (**C, E**) n.gastrocnemius (Arshavsky et al. 1980b)

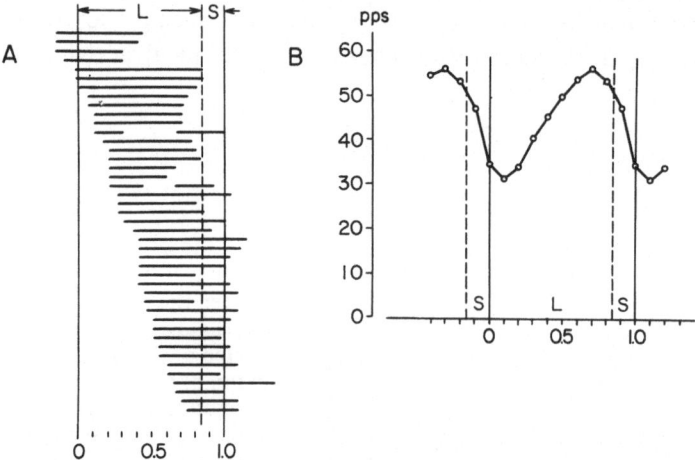

**Fig. 80A, B.** Relationship between the activity of IN neurons and the phase of the cycle during fictitious scratching. **A** The phase distribution of 43 IN neurons. **B** The discharge frequency of the "average" IN neuron as a function of the phase of the cycle (Arshavsky et al. 1980b)

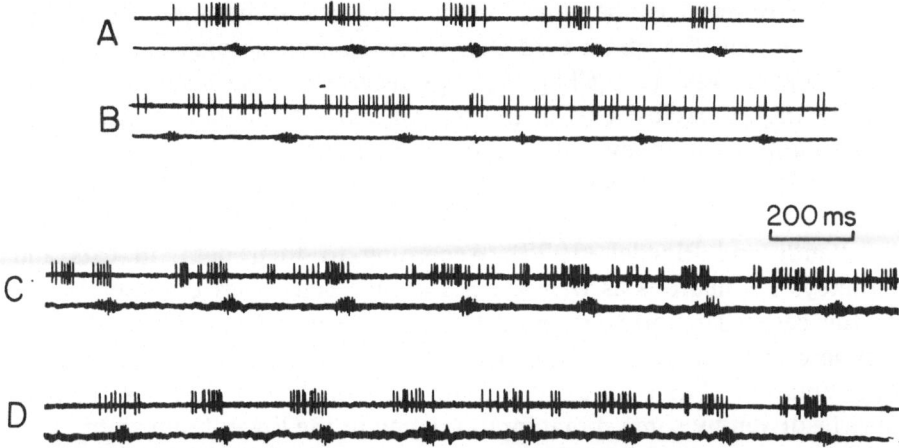

**Fig. 81 A–D.** Effects of contralateral hemisection of the spinal cord and of transection of ipsilateral lateral funiculus upon the activity of IN neurons during fictitious scratching. **A, B** A neuron was recorded before (**A**) and after (**B**) hemisection at the L1 level. **C, D** A neuron was recorded before (**C**) and after (**D**) transection of the ipsilateral lateral funiculus at the C4 level. The *lower trace* is the gastrocnemius ENG (Arshavsky et al. 1980b)

activity of IN neurons (cf. Fig. 81C and D). This finding is in agreement with that of MacKay and Murphy (1974) who found only weak effects produced in the IN by signals coming through the SRCP. However, this finding contradicts the conclusion made by McCrea et al. (1977) that the SRCP efficiently affects the IN neurons both through the collaterals of reticulo-cerebellar fibres and through the Purkinje cells.

### 4. Lateral Nucleus

The lateral cerebellum (the lateral part of the hemisphere and the lateral nucleus) has been greatly developed through mammalian phylogeny becoming the most prominent portion of the cerebellum in Primates (Larsell and Jansen 1972). In contrast, in cats the lateral nucleus (LN) is the most poorly developed of all three cerebellar nuclei; its volume is about 9,9 mm$^3$, it contains approx. 5,800 neurons (Palkovits et al. 1977).

The lateral cerebellum is connected mainly with the cerebrum (see Evarts and Thach 1969; Allen and Tsukahara 1974). It receives signals from different areas of the cerebral cortex via the cerebro-ponto-cerebellar pathway as well as via the cerebro-reticulo-cerebellar and cerebro-olivo-cerebellar pathways.

More obscure is the question of connections between the lateral cerebellum and the spinal cord. According to morphological and electro-physiological data (see Chap. II), spino-cerebellar pathways terminate mainly (if not exclusively) in the medial and intermediate zones of the cerebellum. In monkeys, it has been demonstrated that LN neurons practically do not respond to stimulation of peripheral nerves (Allen et al. 1978; Harvey et al. 1979). However, a different result was obtained by Bantli and Bloedel (1977). According to their data, in cats and monkeys LN neurons as well as the Purkinje cells from the lateral zone of the cerebellar cortex respond to stimulation of spinal afferents. The responses have a rather short latency and persist after intercollicular decerebration. These findings led to the conclusion that LN neurons may be driven by some spinal input relayed in the lower brain stem.

The data concerning efferent connections from the LN are not free from contradictions either. According to the majority of authors, the fibres originating from the LN terminate mainly in the ventrolateral nucleus of the thalamus which, in turn, projects to the cerebral cortex (see Jansen and Brodal 1954; Eccles et al. 1967; Evarts and Thach 1969; Allen and Tsukahara 1974). However, some experiments on cats and monkeys have shown that stimulation of the LN may also evoke reactions in the spinal neurons (Bantli and Bloedel 1976; Schultz et al. 1976, Bloedel and Bantli 1978). These effects persist after removal of the motor cortex. Thus, it was concluded that the LN can influence the spinal cord not only through the cerebral cortex, but also through the brain stem structures. In particular, the LN can evoke monosynaptic excitation of RS neurons (Bantli and Bloedel 1975; Bloedel and Bantli 1978; Tolbert et al. 1980).

To determine whether the rhythmical process in the spinal hindlimb centre affects LN neurons, their activity was recorded during the ficti-tious scratch reflex (Arshavsky et al. 1980b). It was found that the activity of LN neurons, unlike that of neurons of other cerebellar nuclei, was not rhythmically modulated during fictitious scratching (Fig. 82). Only some tonic activation was observed in a number of the neurons.

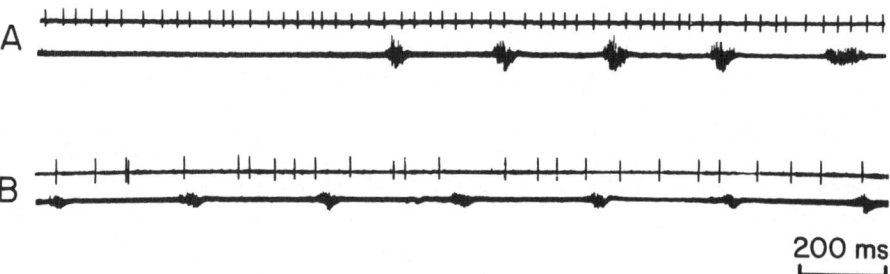

200 ms

Fig. 82A, B. Activity of two LN neurons during fictitious scratching (Arshavsky et al. 1980b)

Similar results were obtained while studying vestibular responses of neurons of cerebellar nuclei (Antziferova et al. 1979a; Arshavsky et al. 1980a). It was found that only FN and IN neurons respond to tilting of the cat in the frontal plane (Fig. 85A) while LN neurons do not. All these results do not agree with the point of view that the LN is directly related to the spinal centres.

The behaviour of LN neurons during scratching differs from that during voluntary movements controlled by the higher centres. In chronic experiments on monkeys (Thach 1968, 1970a, 1975, 1978a, b; Grimm and Rushmer 1974; Robertson and Grimm 1975; Stein J 1978; Strick 1983; Harvey et al. 1979) as well as on rats (Hernandez-Mesa and Bureš 1978), it was demonstrated that such movements are accompanied by changes in the activity of dentate (lateral) nucleus neurons. These data support the idea that the LN participates in the activity of the cerebral, but not spinal motor centres (Evarts and Thach 1969; Allen and Tsukahara 1974, Brooks 1974; Meyer-Lohmann et al. 1977; Beaubaton et al. 1978; Massion and Sasaki 1979; Trouche and Beaubaton 1980; Brooks and Thach 1981; Strick 1983).

## 5. Conclusion

In the present chapter a description of signals at various points of the spino-cerebellar loop has been completed. The signals were studied in more detail for fictitious scratching due to advantages presented by the immobilized preparation. The main results of this study are presented in Figs. 83 and 84 for nervous structures related to the cerebellar vermis and pars intermedia, respectively. The neuron groups studied are shown on the left, the connections were found to be of greater importance for transmitting the rhythmical influences between groups being indicated. The signs of direct (monosynaptic) influences between neurons of various groups, known up to date, are also shown.

Rhythmical signals which determine the rhythmical activity of Purkinje cells reach the cerebellar cortex from the spinal hindlimb centre through the VSCT. There are no essential differences between the signals coming through this tract to the vermis and pars intermedia (see Chap. II). The Purkinje cells from the vermis (Fig. 83) act upon the FN neurons. Besides, the FN neurons receive rhythmical signals from the spinal hindlimb centre through the SRCP; these signals arrive via axon collaterals of the LRN neurons. The FN neurons control the RS neurons. The Purkinje cells from the vermis control not only the FN neurons, but also the VS neurons.

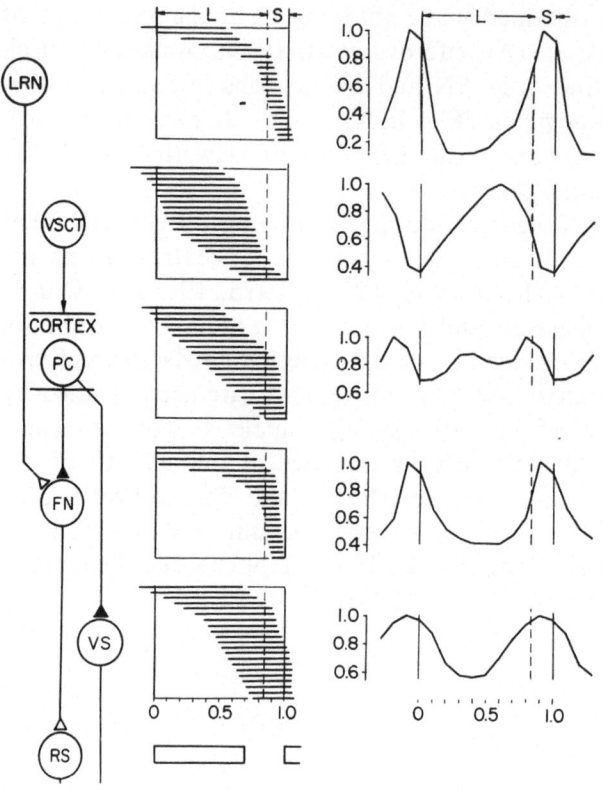

**Fig. 83.** Interaction of neurons related to the cerebellar vermis and their rhythmical activity during the fictitious scratch reflex. See text for explanation

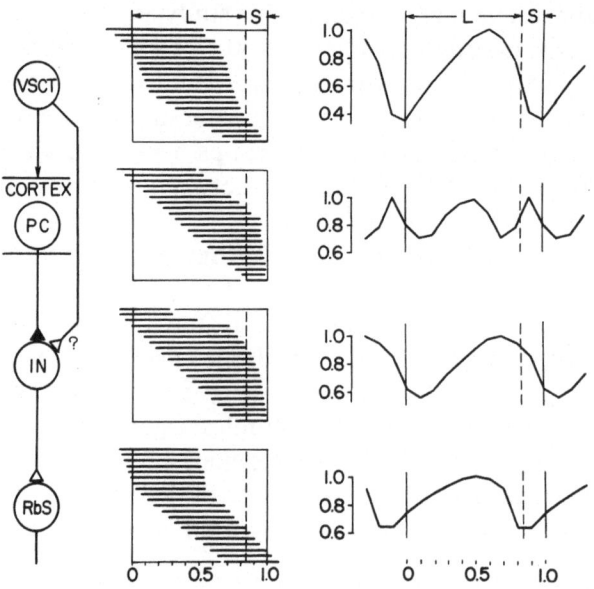

**Fig. 84.** Interaction of neurons related to the pars intermedia of the cerebellum and their rhythmical activity during the fictitious scratch reflex. See text for explanation

The Purkinje cells from the pars intermedia (Fig. 84) control the IN neurons. Besides this input, the IN neurons may receive rhythmical signals through axon collaterals of the VSCT; however, the contribution of this input to the rhythmical activity of IN neurons has not been determined. In turn, the IN neurons control the RbS ones.

Figures 83 and 84 schematically show phase distributions for all the neuron groups studied during fictitious scratching (except for the RS neurons, for which a typical phase of the discharge is shown at the foot of Fig. 83). Besides, on the right, frequency curves characterizing the overall activity of each group of neurons are presented. The curves were obtained from the corresponding curves for "average" neurons by dividing all the ordinates by the maximum one. Let us compare the characteristics of rhythmical activity (i.e., phase distributions and frequency curves) of various neuron groups. One can see from Figs. 83 and 84 that each group contains neurons firing in various phases of the cycle. Furthermore, neither the target neurons for a given neuron are known, nor the cells converging on a given neuron. Thus, until a detailed study of interconnections between individual neurons of the cerebellum and brain stem is performed, a wide range of speculations concerning the interaction between the neurons will persist. The material presented in this book can only restrict, to some extent, the number of possible hypotheses.

*VSCT Neurons and Purkinje Cells.* The population of VSCT neurons is rather homogeneous with respect to their positions in the cycle: most neurons are active in the L-phase. Thus, the overall activity of the VSCT varies considerably in the course of the cycle (by a factor of about 2.5). In contrast, the population of Purkinje cells, both from the vermis and from the pars intermedia, is not homogeneous: the positions of their bursts are almost evenly distributed throughout the cycle and, consequently, the overall activity hardly varies in the course of the cycle.

How can input signals, coming through the VSCT mainly in the L-phase, result in the activation of some Purkinje cells in the L-phase and of other ones in the S-phase? One of the possible explanations is that the VSCT exerts its action upon Purkinje cells in two different ways: it can excite Purkinje cells through granule cells and inhibit them through neurons of the molecular layer (Fig. 13B). Evidence for the significant role of both excitation and inhibition in generating the rhythmical activity of Purkinje cells is provided by the fact that Purkinje cells fire in the bursts at a rate higher than at rest, while in between the bursts they are silent (Fig. 65 A–E).

*Purkinje Cells, LRN and FN Neurons.* Evidence for the existence of two powerful inputs to FN neurons was obtained in experiments with separate interruption of the VSCT and SRCP (see this chapter, Sects.

1 and 2). After the VSCT transection, the rhythmical activity of Purkinje cells disappeared while that of FN neurons only decreased. Transection of the SRCP led to a decrease of the rhythmical activity of FN neurons (Fig. 77A, B) and did not affect the Purkinje cells (Fig. 71C, D). Evidence for the excitatory action of the LRN upon the FN was also obtained from the striking similarity between the phase distributions and between the frequency curves for these two nuclei (Fig. 83) as well as from their similar behaviour during ipsi- and contralateral scratching (Figs. 32 and 74). The inhibitory input to FN neurons must be produced by those Purkinje cells that fire reciprocally to FN neurons. Since most FN neurons fire at the end of the cycle, they must be driven by the Purkinje cells firing at the beginning and in the middle of the cycle. One can see from Fig. 83 that such Purkinje cells constitute about one half of the whole population.

*Purkinje Cells and VS Neurons.* The rhythmical activity of VS neurons, in contrast to that of FN neurons, is determined mainly by the signals coming from the cerebellar cortex. This is proven by the following facts. In Chap. III and IV it was demonstrated that the VSCT affects VS neurons exclusively through the cerebellum since in decerebellate animals rhythmical modulation of the discharge of VS neurons is absent. In Chap. IV and in the present chapter it was demonstrated that the VSCT influences upon VS neurons are not mediated by the FN, but by Purkinje cells. Indeed, the rhythmical activity of both Purkinje cells and VS neurons almost completely disappears after the VSCT transection and hardly changes after the SRCP transection. On the contrary, the SRCP transection strongly affects the FN neurons. Besides, during contralateral scratching, the rhythmical activity of VS neurons is weak. In this respect they resemble the Purkinje cells (see Fig. 69) but not the FN neurons that exhibit a similar pattern of rhythmical activity during both ipsi- and contralateral scratching (Figs. 74–76). Finally, after cerebellar ablation, the firing rate of VS neurons considerably increases, a fact which can be explained by the prevalence of the inhibitory input from Purkinje cells over the excitatory one from the FN.

The question concerning the role of the FN in the control of the VS neurons cannot be considered as completely settled. According to anatomical studies (Brodal et al. 1962), fibres from the FN terminate preferably on small neurons of Deiters' nucleus. One cannot exclude the possibility that such neurons were not recorded in the experiments described above.

The population of VS neurons is not homogeneous; their bursts are distributed throughout the cycle. But the distribution is not even: the units active at the end of the cycle prevail over those active at the beginning and in the middle of the cycle. This is reflected in the frequency curve that has a maximum in the S-phase, the maximum activity being about 1.7 times higher than the minimum one (Fig. 83). Since phases of activity of Purkinje cells are extremely diverse, the pattern of activity of every VS neuron can be explained by the monosynaptic inhibitory action of the reciprocal group of Purkinje cells. In this connection one should note that among the Purkinje cells giving axons to Deiters' nucleus (i.e., responding antidromically to its stimulation, Fig. 65J), the units with all possible phases of activity were found.

Phases of activity of Purkinje cells from the vermis are distributed almost evenly throughout the cycle. On the other hand, in those structures that are under the control of this area of the cerebellar cortex (i.e., in the FN and Deiters' nucleus), the neurons firing at the end of the cycle prevail (Fig. 83). If the FN and VS neurons were driven by Purkinje cells only through well-known monosynaptic inhibitory pathways, the above mentioned fact might mean that a definite part of the population of Purkinje cells (active at the beginning and in the middle of the cycle) plays a more prominent role in the control of VS and FN neurons. But one cannot exclude the possibility that Purkinje cells affect nuclei neurons not only through monosynaptic pathways, but also through polysynaptic ones. In such a case many different explanations for the discrepancy between the phase distribution of Purkinje cells and that of nuclei neurons might exist.

*FN and RS Neurons.* During scratching most RS neurons are inactive (see Chap. III and IV); a few neurons, that are active, fire in the L-phase (this is schematically shown at the foot of Fig. 83). It is not known whether the rhythmical activity of these neurons during scratching is determined by the cerebellum. However, it seems to be the case since the rhythmical activity of RS neurons during locomotion is deter-mined by signals coming from the cerebellum (see Chap. III). If the recorded RS neurons are typical ones (i.e., if they allow assessment of the influences from the FN upon the whole population of RS neurons), then a difficult problem aries with respect to the organization of these influences. Indeed, most FN neurons fire reciprocally to RS ones (Fig. 83). On the other hand, the FN neurons form excitatory synapses on many RS neurons (Ito et al. 1970b; Eccles et al. 1975a). It does not matter that FN neurons give axons both to the ipsilateral and, to a greater extent, to the contralateral reticular formation (Walberg et al. 1962; Eccles et al. 1975a; Batton et al. 1977) since during scratching both fastigial nuclei are active in-phase (Figs. 75 and 76). If only the known monosynaptic excitatory connections between the FN and RS neurons are of importance, it follows that the control of RS neurons is performed by a small portion of FN neurons active in the L-phase. Another hypothesis also seems to be possible, i.e., that the main role in conducting the influences from the FN to RS neurons is not played by mono- but by polysynaptic pathways including inhibitory interneurons. The following fact speaks in favour of such a supposition. When locomotion is evoked, a strong activation of RS neurons is observed both in animals with intact cerebellum and in decerebellate ones (see Chap. III). When stepping movements begin in animals with intact cerebellum, one can see periodical pauses in the original activity (Fig. 43E, F) which can be most easily interpreted as a result of the periodical action of inhibitory neurons. It should be recalled that during locomotion most FN neurons are active in-phase with ipsilateral RS neurons (Figs. 44 and 73). Therefore, FN neurons are active reciprocally to contralateral RS neurons on which they predominantly project.

*IN and RbS Neurons.* We shall not consider relations between the VSCT and Purkinje cells from the pars intermedia since their activities are similar to those of the cells from the vermis. The question of the relations between Purkinje cells and IN neurons is unclear because the contribution of signals conveyed by the collaterals of VSCT fibres to the rhythmical activity of IN neurons has not been appraised.

While considering relations between IN neurons and RbS neurons (Fig. 84), one can see that both cell populations are not homogeneous with respect to the phase of activity in the cycle, but their activity in the L-phase is somewhat higher, on average, than that in the S-phase. However, it is important to note that the portion of cells active at the beginning of the cycle is greater in the red nucleus than in the IN which is also reflected in the corresponding frequency curves. To explain this fact on the basis of known monosynaptic excitatory connections between the IN and red nucleus, one can assume that the control of a large group of RbS neurons active at the beginning of the cycle is performed by a small group of IN neurons. However, one cannot exclude another explanation, i.e., transmission of influences from IN to RbS neurons through polysynaptic pathways, including inhibitory interneurons.

Thus, while considering interactions between various groups of cerebellar neurons as well as between cerebellar and brain stem neurons, one faces some difficulties in attempts to explain the experimental data only on the basis of known monosynaptic connections between various neuron groups. It appears that the interaction between these groups is more complicated.

In conclusion, one more question concerning relations between the cerebellar cortex and the nuclei can be considered. One can see from numerous examples presented in this chapter that the rhythmical activity of neurons of cerebellar nuclei (as well as of Deiters' nucleus, see Chaps. III and IV) is considerably more regular than that of Purkinje cells. This can be explained by the well-known fact that many of

Purkinje cells converge upon each nucleus neuron. However, detailed organization of such a convergence is unknown. One can suppose that a given nucleus neuron receives inputs from the Purkinje cells located in close proximity. The finding that neighbouring Purkinje cells fire in-phase (Fig. 66) speaks in favour of such a hypothesis. Also there are some indications that the Purkinje cells, having similar afferent inputs, are located close to each other, the neighbouring cells projecting to a definite region of the given nucleus ("microzones", see Andersson and Oscarsson 1978b; Ekerot et al. 1979; Giuffrida et al. 1981). One may suppose that phase relations between small groups of neighbouring Purkinje cells and corresponding groups of nuclear neurons differ from those found by averaging the activity of large populations of cells. Evidently, the problem of detailed anatomical and functional organization of cortico-nuclear projections in the cerebellum needs further study.

# VI External Inputs of the Spino-Cerebellar Loop

The preceding chapters have dealt with the signals at different points of the spino-cerebellar loop. Both the neurons of descending tracts and those of the cerebellar cortex are driven by preceding elements of the loop. Besides, they receive signals from various receptors and from other parts of the brain ("external inputs", Fig. 1). For the VS neurons "external inputs" are those from the vestibular receptors and from different vestibular nuclei (Brodal et al. 1962). The RS neurons receive numerous fibres from the cerebral cortex and other parts of the brain (Brodal 1957) including "locomotor regions" of the brain stem (Orlovsky 1970a). The RbS neurons receive inputs mainly from the cerebral cortex (Massion 1967; Eccles et al. 1967). The medial and intermediate part of the cerebellum which sends axons to the neurons of descending tracts, receives signals from the vestibular receptors and vestibular nuclei (Brodal et al. 1962), and from the cerebral cortex as well as from other parts of the brain (Jansen and Brodal 1954; Dow and Moruzzi 1958; Evarts and Thach 1969; Allen and Tsukahara 1974). There is no doubt that these "external inputs" can affect the activity of the corresponding elements of the spino-cerebellar loop.

One of the few "external inputs" to the neurons of the cerebellum and descending tracts which persists after decerebration is that from the vestibular apparatus. The effects of this input upon the VS neurons were studied during locomotion (Orlovsky and Pavlova 1972b, c) and fictitious scratching (Arshavsky et al. 1978d). To stimulate vestibular receptors, the animal was tilted in the frontal plane (Fig. 85A). Such stimulation evokes responses mainly in the otolith receptors (Schor 1974). In a state of rest before locomotion or scratching is evoked, the tilting of the animal results mainly in dynamic responses of VS neurons. In most units (particularly, in the larger ones, see Panchin 1978) the activity increases during the ipsilateral tilt (Fig. 85B). These responses are mediated by the cerebellum since they are considerably reduced in decerebellate animals (Orlovsky and Pavlova 1972a).

The first attempt to demonstrate an interaction between the two inputs (i.e., the "locomotor" and vestibular ones) to VS neurons gave no result. It was discovered that responses of VS neurons to vestibular

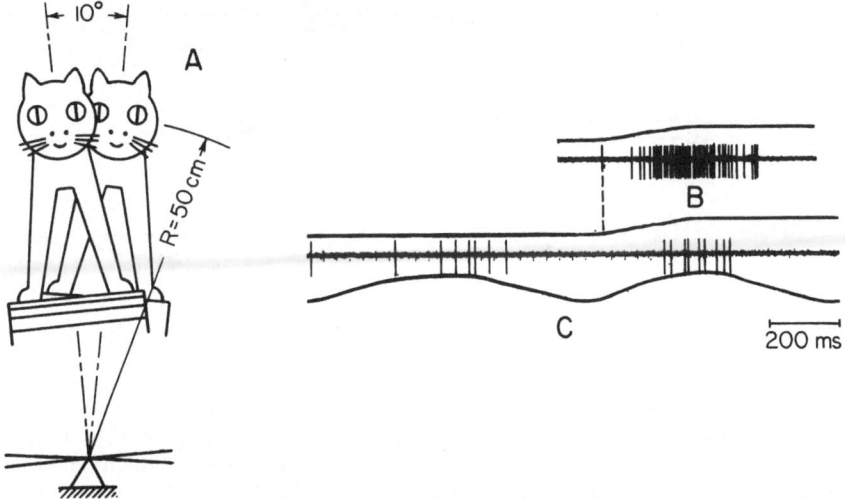

**Fig. 85 A—C.** Suppression of vestibular reactions in a VS neuron during locomotion. A Scheme of the experiment. **B** A response of the neuron to a tilt of the cat before elicitation of locomotion. **C** The neuron does not respond to the tilt during loco-motion. The *upper trace* is the angle of the tilt (ipsilateral, i.e. towards the side of the neuron, tilt — up). The *lower trace* in **C** is the movement of the ipsilteral hindlimb (protraction — up) (Orlovsky and Pavlova 1972b, c)

stimulation are suppressed during locomotion. The vestibular input which was very effective at rest (Fig. 85B) is almost inefficient during locomotion (Fig. 85C). What might be the functional significance of the weakening of vestibular responses of Deiters' neurons during locomotion? One might suppose that the discharges of vestibular receptors in answer to the irregular and accidental head and body movements during rapid running could disturb the rhythmical activity of Deiters' neurons and, thus, prevent the proper stepping performance.

Another result was obtained in the study of vestibular reactions of Deiters' neurons during fictitious scratching (Arshavsky et al. 1978d). In this case, no suppression of vestibular reactions was observed. Figure 86A shows the responses of a VS neuron to periodical tilting under resting conditions, before scratching is evoked: the neuron fires during ipsilateral tilting. When the fictitious scratch reflex is evoked, the neuron exhibits rhythmical bursting (B). This rhythmical activity depends on the vestibular influences (C): while the animal is being tilted to the ipsilateral side (relative to the neuron), rhythmical activity of the neuron, correlated with scratching, considerably increases (i.e., the bursts become longer and the firing rate higher). In contrast, during contralateral tilting, the neuron is silent. Since vestibular reactions of Deiters' neurons persist during scratching, one might suppose that vestibular signals

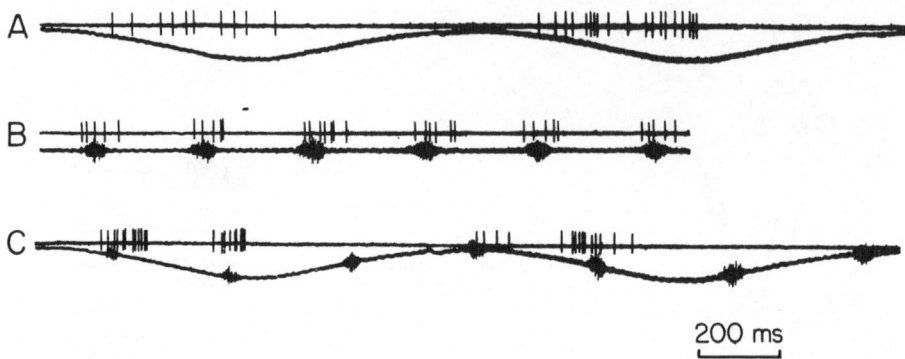

**Fig. 86 A—C.** Vestibular reactions of a VS neuron during fictitious scratching. A Reactions of the neuron to tilts before elicitation of scratching. **B** Activity of the neuron during fictitious scratching. **C** Reactions of the neuron to tilts during fictitious scratching. The *lower trace* is the tilt angle (ipsilateral tilt — down) and the gastrocnemius ENG (Arshavsky et al. 1978d)

help to maintain posture when the animal performs scratching movements. Indeed, the head position in this case is more stable than during locomotion since the hindlimb performing rhythmical movements touches the head only slightly, and vestibular signals are able to monitor the head position.

Thus, the cerebellum can exert two kinds of influences upon VS neurons, determined by signals coming, respectively, from the spinal rhythmical generator and from the vestibular receptors. In the latter case the signals may be considered as "external" in relation to the spino-cerebellar loop. As a result of interaction of two inputs signals conveyed by the VS tract to the spinal cord depend both on the phase of the scratch cycle and on the head position.

The results obtained for the VS tract can be extrapolated to other descending tracts and formulated in the following way. During locomotion and scratching, the excitability of cerebellar and descending tract neurons is rhythmically modulated in relation to the activity of the spinal rhythmical generator. Thus, signals from various receptors and from other brain centres are not free to reach the spinal motor centres: the efficiency of their transmission depends on the phase of rhythmical movements. The excitability of VS neurons is highest during the extensor phase of the cycle. Therefore, "external" (in relation to the spino-cerebellar loop) signals are most efficiently transmitted through the VS tract in the extensor phase of the cycle. This is in accordance with the fact that excitation of extensor motoneurons is the main effect produced by the VS tract in the spinal cord. In contrast, the excitability of RS and RbS neurons has its maximum in the flexor phase of the cycle. Therefore, it is in the flexor phase of the cycle when the RS and RbS tracts, exciting the flexor motoneurons, are most efficient in transmitting "external" signals.

Thus, signals from the spinal rhythmical generators of scratching and locomotion determine rhythmical modulation of the sensitivity of the cerebellar and descending tract neurons to signals arriving through the "external" inputs of the spino-cerebellar loop. We think that the control of transmission of messages from various brain centres to the spinal cord is one of the main cerebellar functions. In the next chapter we shall try to support this point of view.

# VII Role of the Cerebellum in Control of Locomotion and Scratch Reflex

In the Introduction, a definition of the spino-cerebellar loop was given. This notion is convenient for describing the activity of the cerebellum and related structures during locomotion and scratching. One can define the spino-cerebellar loop   (Fig. 1) in the following way. (1) The spinal cord generates scratching or stepping movements. (2) Spino-cerebellar pathways convey to the medial and intermediate parts of the cerebellum signals on the activity of neuronal mechanisms of the spinal cord and on the current state of the executive motor apparatus. (3) On the basis of these signals, the cerebellum generates its output signals which modulate the activity of neurons of descending brain stem-spinal tracts. (4) Signals coming to the spinal cord via the descending tracts affect its activity. (5) The cerebellum as well as the neurons of descending tracts have "external inputs" through which they receive signals from other parts of the brain. In previous chapters, the activity of the main nervous mechanisms participating in the control of locomotion and scratching (i.e., the spinal cord and the cerebellum) as well as the signals that these mechanisms send to each other were described. The influence of one of the "external inputs" (that from the vestibular apparatus) upon the activity of the spino-cerebellar loop was also considered.

In the present chapter we advance a hypothesis concerning the role of the cerebellum in the control of locomotion and scratching, and try to support it by experimental data. This chapter consists of a number of statements, the validity of which we will try to confirm.

*1) Motor Activity of Animals is Based on a Restricted Number of Motor Synergisms (Programmes). Each of Them Can be Put Into Operation by a Relatively Simple Command.* The term "programme" brings to mind associations with a computer, but we do not think that the brain functions according to the same principles as a computer. Thus, we shall use the term "synergism" (see Bernstein 1967), meaning a physiological equivalent of the "programme" though we cannot provide any precise definition of the term. We shall, therefore, give some examples of what we call "synergisms".

Motor synergisms can be perfectly exemplified by the nervous mechanisms of locomotion and scratching to which the present book is devoted. Locomotion is evoked by electrical stimulation of the locomotor regions of the midbrain or subthalamus. This stimulation triggers the spinal mechanism which produces stepping movements of all four limbs, their coordination as well as (to some extent) their adaptation to external conditions (for example, stepping over an obstacle). Scratching is evoked by electrical stimulation of the spinal cord or by natural stimulation of the receptive field of the scratch reflex. In the latter case, the spinal mechanism provides scratching of the very point which is irritated.

Thus, a complex motor act is triggered by a simple command. The programme of the motor act is determined by the structure of the control mechanism (i.e., of the synergism). Some more examples of motor synergisms are the nervous mechanisms of respiration, mastication, swallowing, jumping, standing, catching the prey, etc. These synergisms are located in different parts of the nervous system: swallowing and mastication are controlled by the medulla oblongata, while in catching the prey the cerebral cortex must necessarily be involved. The nervous mechanisms of locomotion and scratching are spinal synergisms.

*2) The Cerebellum Receives Detailed Information on the Activity of Motor Synergisms.* The rhythmical signals related to the hindlimb movements reach the medial and intermediate parts of the cerebellum via three spino-cerebellar pathways: DSCT, VSCT and SRCP (see Chap. II). The DSCT conveys detailed information concerning the activity of the executive motor apparatus: the phase and degree of contraction of various muscles, joint angles, and contact with the ground are the data monitored by this tract. These signals show the results of the activity of spinal nervous mechanisms.

The VSCT and SRCP convey completely different information, namely, information about the activity of neuronal mechanisms of the spinal cord (the rhythmical generators) that determine the main characteristics of movements — the rhythm and duration of the flexor and extensor phases of the cycle. Comparison of the activity of VSCT and SRCP neurons with that of spinal interneurons from the "leading" region of the hindlimb centre (see Chap. II) has revealed that practically all types of spinal interneurons of this region have their "counterparts" among the neurons of spino-cerebellar pathways. One may speculate that both the VSCT and the SRCP convey information on the activity of different groups of interneurons of the "leading" region of the lumbosacral spinal cord.

Thus, during locomotion and scratching, the cerebellum receives information on the activity of the spinal neuron mechanism (rhythmical generator) that determines the main characteristics of movement (via the VSCT and SRCP), and information on the activity of the executive motor apparatus (via the DSCT).

*3) Information Received by the Cerebellum is Transformed in Order to Provide the Essential Aspects of the Activity of Motor Synergisms.* As a rule, motor synergisms do not work in isolation, but in cooperation with one another. For instance, during locomotion, the animal does not only have to perform rhythmical limb movements and exercise interlimb coordination (locomotor synergism), but must also maintain equilibrium (synergism of "standing"). While hunting, the animal must combine locomotor movements with those of catching the prey, etc. Furthermore, any synergism must be adapted to changing external conditions. This question will be given more detailed consideration later. Here, we just want to draw the reader's attention to the fact that when dealing with the problem of cooperation of several synergisms, one must possess certain data concerning the current state of each of them; the same being true with regards to interrelations between synergisms and environment. We think that these problems might be solved much more efficiently when instead of the full information on the state of synergisms and environment only a relatively small part of it, providing the most *relevant* data, is taken into account. We believe that it is the cerebellum that selects these data out of the immense inflow of information.

Signals coming to the spinal cord from the brain stem via the descending tracts (VS, RS and RbS), which can be regarded as the result of action of the cerebellar output signals upon the neurons of these tracts, differ greatly from the signals received by the cerebellum from the spinal cord. The cerebellar input signals contain detailed information on the activity of the spinal rhythmical generator and on the current state of the executive motor apparatus. In the signals reaching the spinal cord this information is considerably reduced.

The statement that information on the current state of the executive motor apparatus is reduced, is based on the following facts (see Chap. IV). The activity of neurons of descending tracts is almost the same in both actual and fictitious scratching. On the other hand, during fictitious scratching the signals on the activity of the executive motor apparatus are absent since the latter does not function. Thus, these signals are not used for generating the cerebellar output signals during scratching. Nor are they used during locomotion. This was proven in the experiments with artificial disturbances of stepping movements. Considerable perturbations of movements at the knee and ankle joints (temporary arrest of the joint or acceleration of its flexion) sarcely affect the activity of RbS neurons, the neurons most sensitive (among the neurons of descending tracts) to afferent signals coming via the DSCT. One can imagine that the influence of such disturbances upon VS and RS neurons will be still smaller.

The signals transmitted to the cerebellum by the VSCT and SRCP are themselves, to a great extent, of an integral nature. Indeed, they carry information on the activity of the rhythmical generator only, i.e., signals from the mechanism that determines the main characteristics, but not the details of movements. Nevertheless, this information is further reduced in the cerebellum. The signals conveyed to the spinal cord via descending tracts do not contain information concerning the activity of different groups of spinal interneurons, though such information is present in the input signals. Unlike the VSCT and SRCP neurons, the neurons of descending tracts do not "copy" the activity of different groups of spinal interneurons. Thus, observing the signals conveyed by descending tracts, one cannot determine the activity of different groups of spinal interneurons participating in the generation of rhythmical oscillations.

What, then, is the content of the cerebellar output signals? Observing the activity of VS, RS and RbS neurons during locomotion and scratching, one can argue that there is a rhythmical process in the spinal cord, and determine, at least, the intensity of this process and its frequency. One can also determine which of the four limbs is involved in rhythmical movements, and the duration of the flexor and extensor phases of the cycle. Finally, the phase of the movement of each limb at a given moment can be established.

We believe that having information on the current state of the spinal level of motor control at is disposal, the cerebellum can select the specific data, depending on the character of the motor task which is to be solved by the animal. The present study deals with rhythmical movements; and it was found that the cerebellum singles out the essential characteristics which can provide one with enough information on the rhythmical activity of the spinal synergism. One can suppose that studies of other synergisms would reveal that the cerebellum singles out their main characteristics as well. In this respect, the data obtained by Robertson and Grimm (1975) are very interesting. They recorded the activity of dentate neurons in Cebus monkeys trained to touch three horizontally positioned buttons in a left-to-right sequence. By changing the position of the button panel in relation to the monkey, different joint angles and different sets of muscles as well as changes in the timing of the performance were required to perform the task. It was found that the discharge patterns of the dentate neurons during the performance of the motor task did not depend on the concrete trajectories of the movements. This means that the activity of dentate neurons was related to some general aspect of the motor performance, but not specifically to the actions of the joints or muscles used to execute the task.

Apparently, the cerebellum also performs the analysis of signals coming from sense organs and selects essential data on the environment. In this connection, one should mention the results obtained while studying the reactions of cerebellar neurons to binaural acoustic stimuli (Altman et al. 1976; Bechterev 1978). It was determined that in the cerebellar acoustic area (lobules VI and VII), reactions of neurons to acoustic stimuli depend strongly on the spatial localization of the source of sound or on the direction of its motion. On the other hand, reactions of the cerebellar neurons, unlike those of the auditory system, hardly depend on such parameters of the stimulus as intensity, duration and frequency of sound.

Apparently, the morphology of the cerebellum promotes the selection of substantial information from the immense inflow of afferent signals. Intracerebellar connections are characterized by high convergence and divergence (see Chap. II and V). This might account for the fact that many "details" contained in the input signals are excluded in each stage of their passing through the cerebellum. Nevertheless, the experiments described in this book have shown that intracerebellar connections are far from being chaotic since only the details are left out while all relevant information is preserved and goes into the cerebellar output.

*4) The Cerebellum Neither Belongs to Motor Synergisms Nor Participates in Their Triggering.* This statement is based on the fact that spinal animals are capable of performing stepping and scratching movements (see Chap. I), and cerebellotomy does not prevent triggering of the spinal synergisms of these movements. Thus, the spino-cerebellar loop as such is not necessary for generating locomotor and scratching movements.

Numerous studies on the effects of cerebellar ablation have demonstrated that the cerebellum is unnecessary for spinal synergisms and also for many other synergisms with, probably, only one exception. In phylogeny, the cerebellum initially developed in close relation with the vestibular apparatus, i.e., as an organ directly participating in the control of equilibrium (Larsell 1967; Llinas 1969). This cerebellar function seems to remain in the animals with a highly developed cerebellum as well.

*5) The Cerebellum Regulates the Transmission of Signals from Various Motor Brain Centres to the Spinal Cord.* Different motor centres of the brain (cerebral cortex, basal ganglia, hypothalamus, reticular formation, etc.) can affect the spinal cord only in one way — by sending signals via the descending tracts. These motor centres can influence the neurons of descending tracts either directly or through the cerebellum. The cerebellum is known to receive signals from all the motor centres as well as from various receptors (see Jansen and Brodal 1954; Dow and Moruzzi 1958; Evarts and Thach 1969; Arshavsky 1972; Allen and Tsukahara 1974). In turn, the cerebellum can affect the neurons of all the main descending tracts. Thus, the cerebellum is quite capable of regulating the transmission of signals from various brain centres to the spinal cord.

During locomotion and scratching, the signals from the spinal cord evoke rhythmical activity in the cerebellar neurons (see Chap. V) which, in turn, perform rhythmical modulation of the neurons of descending tracts (see Chap. III and IV). As a result, the sensitivity of both the cerebellar and descending tract neurons to signals arriving through "external inputs" of the spino-cerebellar loop (Fig. 1) from other brain centres and from various receptors is rhythmically changing in relation to the activity of spinal mechanisms.

Thus, during locomotion and scratching, the cerebellum rhythmically regulates the transmission of signals from different brain centres and receptors to the spinal cord. This was experimentally proven for the signals coming from the vestibular apparatus. During scratching, the vestibular influences upon the spinal cord transmitted via the VS tract were found to be rhythmically modulated (see Chap. VI).

*6) The Cerebellum Coordinates Different Motor Synergisms and Adapts Them to the Environment.* When a synergism is put into operation, it has three main functions: (1) generation of the movement, in the present case — locomotion or scratching; (2) adaptation of the movement to the environment; (3) cooperation with other synergisms. The first function, i.e., generation of locomotor or scratching movements by the spinal cord, was given detailed consideration in Chap. I. Here, we shall consider the question of cooperation between a given synergism and other ones, and its interaction with the environment. This question is very important since it is evident that an animal that can only choose among a set of separate fixed synergisms would be helpless in the constantly changing real world where it often has to perform several actions at a time.

Some problems of adaptation to the environment are being solved already at the spinal level. As it was mentioned in Chap. I, during locomotion the spinal animal is able to step over an obstacle. However, in most motor tasks it is necessary to consider the spatial position of the body (i.e., the position in the gravity field and in relation to external objects). Such problems can only be solved provided the brain is involved, together with its appropriate receptors (vestibular, acoustic and visual) as well as with centres capable of analyzing the information coming from these receptors.

We think that the problems of interaction between a given synergism and the environment as well as those of cooperation between different synergisms can be solved on the same basis. For instance, both problems can be considerably simplified if detailed information on the current state of the synergisms and the environment is not used, but rather a restricted amount of essential data. Some examples of such data, for locomotion and scratching, were given in Sec. 3 of this chapter. They are: duration of the cycle, phase of the movement, its amplitude, etc. Examples of essential information on the environment were given in the same section: the location of the source of sound or the direction of its movement. The concrete types of cooperation between different synergisms as well as of interaction between a synergism and the environment seem to be highly diverse. Unfortunately, their study requires experimental conditions much closer to natural ones than those of our experiments. We shall, therefore, only try to illustrate some of the types by a few simple examples.

Let us consider the cooperation between locomotion and some other motor synergisms. Suppose a running cat comes across an obstacle it has to jump over. To manage this without stopping, the nervous system has to send an additional excitation to the limb extensors just at the moment when the limb touches the ground, i.e., in the stance phase. Similarly, to step over an obstacle, the flexor activity has to be increased in the swing phase which would lead to an increase of limb flexion. In both examples, the locomotor synergism is the dominating one: in the first case over the synergism of jumping, in the second one over the synergism of stepping over an obstacle. The cooperation between synergisms does not disturb the temporal pattern of the dominating movement. Obviously, in locomotion, the constancy of the temporal pattern of the muscle activities in the cycle is very important both for interlimb coordination and for the maintenance of equilibrium. If, for instance, in diagonal locomotion (i.e., walking and trotting when the diagonal limbs are moving in-phase and the symmetrical ones reciprocally) one of the limbs begins the swing before the symmetrical one touches the ground, the animal will fall down.

In the above described cases, external influence upon the loco-motor mechanism has to be correlated with a particular phase of the locomotor cycle. In other words, the cooperation between synergisms requires the knowledge of their current state. The same examples show that detailed information on the current state of a synergism is unneces-sary; a few essential characteristics (or even one of them — the phase of the cycle as in the above examples) are enough.

What kind of a nervous mechanism could provide the cooperation between synergisms and the interactions between a synergism and the environment? The synergisms as well as the mechanisms analyzing the environment may be located in different parts of the nervous system (for instance, the locomotor and scratching synergisms are located in the spinal cord, while the synergism of jumping over an obstacle seems to be located in the cerebral cortex). The cooperation between "remote" synergisms could be provided by "representatives" ("receptors") of all other synergisms in the domain of the given one. But such a system is vulnerable from an evolutionary point of view since the development of synergisms, the increase of their number and of the complexity of their combinations, would have made the system overcomplicated. We suppose that the problem of cooperation could be solved much more easily if there was a special organ responsible for the cooperation between the synergisms and for the interaction between a synergism and the environ-ment. The cerebellum meets all the requirements for such a role because: (1) it receives detailed information on the state of motor synergisms and of the environment; (2) from this information it selects essential data concerning both the activity of motor synergisms and the state of the environment; (3) it can efficiently regulate the transmission of signals from one part of the nervous system to the other. Apparently, the medial and intermediate cerebellum, considered in the present book, deals with the activity of the spinal and brain stem synergisms, while the lat-eral cerebellum is concerned with that of the synergisms of a higher level.

The prominent role of the cerebellum in organizing the interaction between synergisms is also proven by the fact that one of the typical symptoms of the cerebellar disfunction is "asynergia", i.e., a disturbance of normal interaction between various nervous centres controlling complicated movements (Babinski 1899; Holmes 1939; Dow and Moruz-zi 1958).

In conclusion, we would like once more to draw attention to the amazing fact that the cerebellum receives information from all the motor centres and from the majority of receptors and, in turn, sends the signals to all the motor centres. At the same time, the cerebellum is unnecessary for any particular movement. We consider this phenomenon less surprising if one accepts the hypothesis that the cerebellum provides cooperation between the synergisms and interaction between a synergism and the environment.

# References

Abzug C, Maeda M, Peterson BW, Wilson VJ (1974) Cervical branching of lumbar vestibulospinal axons. J Physiol (Lond) 243:499–522

Adrian ED (1943) Afferent areas in the cerebellum connected with the limbs. Brain 66:289–315

Afelt Z, Veber NV, Maksimova EV (1973) Reflex activity of chronically isolated spinal cord of the cat (in Russian). Nauka, Moscow

Akaike T, Fanardjian VV, Ito M, Nakajima H (1973) Cerebellar control of the vestibulospinal tract cells in rabbit. Exp Brain Res 18:446–463

Allen GI, Tsukahara N (1974) Cerebrocerebellar communication systems. Physiol Rev 54:957–1006

Allen GI, Sabah NH, Toyama K (1972a) Synaptic actions of peripheral nerve impulses upon Deiters neurones via the climbing fibre afferents. J Physiol (Lond) 226: 311–333

Allen GI, Sabah NH, Toyama K (1972b) Synaptic actions of peripheral nerve impulses upon Deiters neurones via the mossy fibre afferents. J Physiol (Lond) 226: 335–351

Allen GI, Gilbert PFC, Marini R, Schultz W, Yin TCT (1977) Integration of cerebral and peripheral inputs by interpositus neurons in monkey. Exp Brain Res 27: 81–99

Allen GI, Gilbert PFC, Yin TCT (1978) Convergence of cerebral inputs onto dentate neurons in monkey. Exp Brain Res 32:151–170

Allen WF (1924) Distribution of fibers originating from the different basal cerebellar nuclei. J Comp Neurol 36:399–439

Alstermark B, Lindström S, Lundberg A, Sybirska E (1981) Integration in descending motor pathways controlling the forelimb in the cat. 8. Ascending projection to the lateral reticular nucleus from C3–C4 propriospinal neurons also projecting to forelimb motoneurons. Exp Brain Res 42:282–298

Altman JA, Bechterev NN, Radionova EA, Shmigidina GN, Syka J (1976) Electrical responses of the auditory area of the cerebellar cortex to acoustic stimulation. Exp Brain Res 26:285–298

Amatuni AS (1981) Effects of peripheral nerves, lateral reticular nucleus and inferior olive stimulation on fastigial neurons of the cat cerebellum. (in Russian) Neirofiziologya 13:168–178

Anden NE, Jukes MGM, Lundberg A, Viklicky L (1966) The effect of DOPA on the spinal cord. I. Influence on transmission from primary afferents. Acta Physiol Scand 67:373–386

Andersen P, Eccles JC, Voorhoeve PE (1964) Postsynaptic inhibition of cerebellar Purkinje cells. J Neurophysiol 27:1138–1153

Anderson ME (1971). Cerebellar and cerebral inputs to physiologically identified efferent cell groups in the red nucleus of the cat. Brain Res 30:49–66

Andersson G, Oscarsson O (1978a) Projections to lateral vestibular nucleus from cerebellar climbing fiber zones. Exp Brain Res 32:549–564

Andersson G, Oscarsson O (1978b) Climbing fiber microzones in cerebellar vermis and their projection to different groups of cells in the lateral vestibular nucleus. Exp Brain Res 32:565–579

Andersson G, Sjölund B (1978) The ventral spino-olivocerebellar system in the cat. IV. Spinal transmission after administration of clonidine and l-DOPA. Exp Brain Res 33:227–240

Andersson O, Grillner S, Lindquist M, Zomlefer M (1978) Peripheral control of the spinal pattern generators for locomotion in cat. Brain Res 150:625–630

Angaut P, Bowsher D (1965) Cerebellorubral connexions in the cat. Nature 208: 1002–1003

Angaut P, Bowsher D (1970) Ascending projections of the medial cerebellar (fastigial) nucleus: an experimental study in the cat. Brain Res 24:49–68

Antziferova LI, Arshavsky YuI, Orlovsky GN, Pavlova GA (1979a) Vestibular responses of cerebellar fastigial neurons (in Russian). Neirofiziologya 11:379–381

Antziferova LI, Arshavsky YuI, Orlovsky GN, Pavlova GA (1979b) Activity of neurons of the cerebellar fastigial nucleus during "fictitious scratch-reflex" in the cat (in Russian). Neirofiziologya 11:604–606

Antziferova LI, Arshavsky YuI, Orlovsky GN, Pavlova GA (1980) Activity of neurons of cerebellar nuclei during fictitious scratch reflex in the cat. I. The fastigial nucleus. Brain Res 200:239–248

Appelberg B (1962) The effect of electrical stimulation in nucleus ruber on the response to stretch in primary and secondary muscle spindle afferents. Acta Physiol Scand 56:140–151

Appelberg B, Kosary IZ (1963) Excitation of flexor fusimotor neurons by electrical stimulation in the red nucleus. Acta Physiol Scand 59:445–453

Appelberg B, Jeneskog T, Johansson H (1975) Rubrospinal control of static and dynamic fusimotor neurones. Acta Physiol Scand 95:431–440

Arduini A, Pompeiano O (1957) Microelectrode analysis of units of the rostral portion of the nucleus fastigii. Arch Ital Biol 95:56–70

Armstrong DM (1974) Functional significance of connections of the inferior olive. Physiol Rev 54:358–417

Armstrong DM (1978) The mammalian cerebellum and its contribution to movement control. In: Porter R (ed) Rev Physiol, vol 17, Neurophysiology III, University Park Press, Baltimore, pp 239–294

Armstrong DM, Eccles JC, Harvey RJ, Matthews PBC (1968) Responses in the dorsal accessory olive of the cat to stimulation of hind limb afferents. J Physiol (Lond) 194:125–145

Armstrong DM, Harvey RJ, Schild RF (1971) Climbing fibre pathways from the forelimbs to the paramedian lobule of the cerebellum. Brain Res 25:199–202

Arshavsky YuI (1972) Organization of afferent inputs to the cerebellar cortex (in Russian). Us fiziol nauk 3:24–53

Arshavsky YuI, Berkinblit MB, Fukson OI (1969a) The layer analysis of evoked potentials in the paramedian lobe of the cerebellum (in Russian). Fiziol Zh SSSR 55:429–436

Arshavsky YuI, Berkinblit MB, Gelfand IM, Yacobson VS (1969b) Two types of granule cells in the cerebellar cortex (in Russian). Neirofiziologya 1:167–176

Arshavsky YuI, Berkinblit MB, Gelfand IM, Fukson OI (1970) Peculiarities of influences from lateral reticular nucleus on the cerebellar cortex (in Russian). Neirofiziologya 2:581–586

Arshavsky YuI, Berkinblit MB, Gelfand IM, Fukson OI (1971a) Organization of somatic nerve projections in various areas of the cerebellar cortex in the cat (in Russian). Neirofiziologya 3:166–174

Arshavsky YuI, Berkinblit MB, Gelfand IM, Keder-Stepanova IA, Smeljanskaya EM, Yacobson VS (1971b) The background activity of Purkinje cells in the intact and deafferented cerebellar cortex of the cat (in Russian). Biofizika 16:684–691

Arshavsky YuI, Berkinblit MB, Fukson OI, Gelfand IM, Orlovsky GN (1972a) Recordings of neurones of the dorsal spinocerebellar tract during evoked locomotion. Brain Res 43:272–275

Arshavsky YuI, Berkinblit MB, Fukson OI, Gelfand IM, Orlovsky GN (1972b)
Origin of modulation in neurones of the ventral spinocerebellar tract during
locomotion. Brain Res 43:276–279

Arshavsky YuI, Berkinblit MB, Gelfand IM, Fukson OI, Yacobson VS (1972c)
Suppression of the Purkinje cell reactions by a reticulo-cerebellar volley (in
Russian). Fiziol Zh SSSR 58:208–214

Arshavsky YuI, Berkinblit MB, Gelfand IM, Orlovsky GN, Fukson OI (1972d)
Activity of the neurons of the dorsal spino-cerebellar tract during locomotion.
Biophysics 17:506–514

Arshavsky YuI, Berkinblit MB, Gelfand IM, Orlovsky GN, Fukson OI (1972e)
Activity of the neurons of the ventral spino-cerebellar tract during locomotion.
Biophysics 17:926–935

Arshavsky YuI, Berkinblit MB, Gelfand IM, Orlovsky GN, Fukson OI (1972f) Activity
of the neurones of the ventral spino-cerebellar tract during locomotion of cats
with deafferented hindlimbs. Biophysics 17:1169–1176

Arshavsky YuI, Berkinblit MB, Gelfand IM, Orlovsky GN, Fukson OI (1974) Differences in the performans of spinocerebellar tract neurones during artificial
stimulation and locomotion. In: Kostyuk PG (ed) Mechanisms of Neuronal
Integration in Nervous Center (in Russian). Leningrad, Nauka pp 99–105

Arshavsky YuI, Gelfand IM, Orlovsky GN, Pavlova GA (1975a) Activity of neurones
of the ventral spinocerebellar tract during "fictive scratching". Biophysics
20:748–749

Arshavsky YuI, Gelfand IM, Orlovsky GN, Pavlova GA (1975b) Origin of modulation
in vestibulospinal neurons during scratching. Biophysics 20:946–947

Arshavsky YuI, Gelfand IM, Orlovsky GN, Pavlova GA (1977) Activity of neurones
of the lateral reticular nucleus during scratching. Biophysics 22:177–179

Arshavsky YuI, Gelfand IM, Orlovsky GN, Pavlova GA (1978a) Messages conveyed
by spinocerebellar pathways during scratching in the cat. I. Activity of neurons
of the lateral reticular nucleus. Brain Res 151:479–491

Arshavsky YuI, Gelfand IM, Orlovsky GN, Pavlova GA (1978b) Messages conveyed
by spinocerebellar pathways during scratching in the cat. II. Activity of neurones
of the ventral spinocerebellar tract. Brain Res 151:493–506

Arshavsky YuI, Gelfand IM, Orlovsky GN, Pavlova GA (1978c) Messages conveyed
by descending tracts during scratching in the cat. I. Activity of vestibulospinal
neurons. Brain Res 159:99–110

Arshavsky YuI, Orlovsky GN, Panchin YuV (1978d) Responses of Deiters' neurons
to tilt during scratching (in Russian). Neirofiziologia 10:316–318

Arshavsky YuI, Orlovsky GN, Pavlova GA, Perret C (1978e) Messages conveyed by
descending tracts during scratching in the cat. II. Activity of rubrospinal neurons.
Brain Res 159:111–123

Arshavsky YuI, Orlovsky GN, Pavlova GA (1980a) Vestibular responses of interpositus and dentate neurons of the cerebellum (in Russian). Neirofiziologya
12:93–95

Arshavsky YuI, Orlovsky GN, Pavlova GA, Perret C (1980b) Activity of neurons of
cerebellar nuclei during fictitious scratch reflex in the cat. II. The interpositus
and lateral nuclei. Brain Res 200:249–258

Arshavsky YuI, Gelfand IM, Orlovsky GN, Pavlova GA, Popova LB (1984a) Origin
of signals conveyed by the ventral spino-cerebellar tract and spino-reticulo-
cerebellar pathway. Exp Brain Res 54:426–431

Arshavsky YuI, Orlovsky GN, Popova LB (1984b) Activity of cerebellar Purkinje
cells during fictitious scratch reflex in the cat. Brain Res 290:33–41

Arshavsky YuI, Meizerov ES, Orlovsky GN, Pavlova GA, Popova LB (1985) Activity
of C3–C4 propriospinal neurons during forelimb "fictitious locomotion" in the
cat (in Russian). Neirofiziologya 17:320–326

Asanuma H (1975) Recent development in the study of the columnar arrangement of
neurons within the motor cortex. Physiol Rev 55:143–156

Asanuma H, Hunsperger RW (1975) Functional significance of projection from the cerebellar nuclei to the motor cortex in the cat. Brain Res 98:73—92

Babinski J (1899) De l'asynergie cérébelleuse. Rev Neurol 7:806—816

Bach LMN, Magoun HW (1947) The vestibular nuclei as an excitatory mechanism for the cord. J Neurophysiol 10:331—337

Baldissera F, Bruggencate G ten (1976) Rubrospinal effects on ventral spinocerebellar tract neurones. Acta Physiol Scand 96:233—249

Baldissera F, Roberts WJ (1975) Effects on the ventral spinocerebellar tract neurones from Deiters' nucleus and the medial longitudinal fascicle in the cat. Acta Physiol Scand 93:228—249

Baldissera F, Roberts WJ (1976) Effects from the vestibulospinal tract on transmission from primary afferents to ventral spino-cerebellar tract neurones. Acta Physiol Scand 96:217—232

Bantli H, Bloedel JR (1975) Monosynaptic activation of a direct reticulo-spinal pathway by the dentate nucleus. Pflügers Archiv Gesamte Physiol 357:237—242

Bantli H, Bloedel JR (1976) Characteristics of the output from the dentate nucleus to spinal neurons via pathways which do not involve the primary sensorimotor cortex. Exp Brain Res 25:199—220

Bantli H, Bloedel JR (1977) Spinal input to the lateral cerebellum mediated by infratentorial structures. Neuroscience 2:555—568

Batini C, Moruzzi G, Pompeiano O (1957) Cerebellar release phenomena. Arch Ital Biol 95:71—95

Batton RR, Jayaraman A, Ruggiero D, Carpenter MB (1977) Fastigial efferent projections in the monkey: An autoradiographic study. J Comp Neurol 174: 281—305

Baumgarten R (1956) Koordinationsformen einzelner Ganglienzellen der rhombenzephalen Atemzentren. Pflügers Archiv Gesamte Physiol 262:573—594

Bauswein E, Kolb FP, Rubia FJ (1980) Discharge of cerebellar units during active and passive movements of the awake Rhesus monkey. Neurosci Lett 19, Suppl N 5, p 438

Bayev KV, Kostyuk PG (1972) Investigation of the modes of connection of cortico- and rubrospinal tracts with neuronal elements of the cervical spinal cord in the cat (in Russian). Neirofiziologya 4:158—167

Beaubaton D, Trouche E, Amato G, Grangetto A (1978) Dentate cooling in monkeys performing a visuo-motor pointing task. Neurosci Lett 8:225—229

Bechterev NN (1978) Unit activity of the cat cerebellar tuber vermis in response to binaural pure tone stimuli (in Russian). Fiziol Zh SSSR 64:1398—1405;

Bergmans J, Grillner S (1968) Monosynaptic control of static γ-mononeurones from the lower brain stem. Experientia (Basel) 24:146—147

Berkinblit MB, Deliagina TG, Orlovsky GN, Feldman AG (1977) Activity of pro- priospinal neurons during scratch reflex in the cat (in Russian). Neirofiziologya 9:504—511

Berkinblit MG, Deliagina TG, Feldman AG, Gelfand IM, Orlovsky GN (1978a) Generation of scratching. I. Activity of spinal interneurons during scratching. J Neurophysiol 41:1040—1057

Berkinblit MB, Deliagina TG, Feldman AG, Gelfand IM, Orlovsky GN (1978b) Generation of scratching. II. Nonregular regimes of generation. J Neurophysiol 41:1058—1069

Berkinblit MB, Deliagina TG, Orlovsky GN, Feldman AG (1980) Activity of moto- neurons during fictitious scratch reflex in the cat. Brain Res 193:427—438

Bernstein N (1967) The coordination and regulation of movements. Pergamon, Oxford

Bezhenaru IS (1971) Investigation of brain stem neuronal structures producing early discharges in reticulo-spinal pathways (in Russian). Neirofiziologya 3: 274—283

Bioulac B, Lamarre Y (1977) Etude de certains aspects de la régulation du mouvement volontaire chez le singe. Bordeaux Médical 10:2009–2019

Blessing WW, Goodchild AK, Dampney RAL, Chalmers JP (1981) Cell groups in the lower brain stem of the rabbit projecting to the spinal cord, with special reference to catecholamine-containing neurons. Brain Res 221:35–55

Bloedel JR, Bantli H (1978) A spinal action of the dentate nucleus mediated by descending systems originating in the brain stem. Brain Res 153:602–607

Bloom FE, Hoffer BJ, Siggins GR (1971) Studies on norepinephrine-containing afferents to Purkinje cells of rat cerebellum. I. Localization of the fibers and their synapses. Brain Res 25:501–521

Bowsher D, Westman J (1970) The gigantocellular reticular region and its spinal afferents: a light and electron microscope study in the cat. J Anat (Lond) 106:23–36

Boylls CC (1975) A theory of cerebellar function with application to locomotion. II. The relation of anterior lobe climbing fiber functions to locomotor behaviour in the cat. COINS Technical Report 76–1, Computer and Information Science. University of Massachusetts

Boylls CC (1977) Olivary unit activity and effect of microstimulation during locomotion. Neurosi Abstr 3:55

Bradley GW, Euler C v, Marttila I, Roos B (1975) A model of the central and reflex inhibition of inspiration in the cat. Biol Cybern 19:105–116

Brodal A (1949) Spinal afferents to the lateral reticular nucleus of the medulla oblongata in the cat. J Comp Neurol 91:259–295

Brodal A (1957) The reticular formation of the brain stem. Anatomical aspects and functional correlations. Oliver and Boyd, Edinburgh

Brodal A (1969) Neurological anatomy in relation to clinical medicine. Oxford University Press, New York

Brodal A (1974) Anatomy of the vestibular nuclei and their connections. In: Kornhuber HH (ed) Handbook of sensory physiology, VI/1. Vestibular system, Part I: Basic mechanisms. Springer, Berlin Heidelberg New York, pp 239–352

Brodal A, Pompeiano O, Walberg F (1962) The vestibular nuclei and their connections. Anatomy and functional correlations. Oliver and Boyd, Edinburgh

Brodal P, Marsala J, Brodal A (1967) The cerebral cortical projection to the lateral reticular nucleus in the cat, with special reference to the sensorimotor cortical areas. Brain Res 6:252–274

Brooks VB (1974) Some examples of programmed limb movements. In: Motor aspects of behaviour and programmed nervous activities. Brain Res 71:299–308

Brooks VB, Thach WT (1981) Cerebellar control of posture and movement. In: Brooks VB (ed) Handbook of physiology. The nervous system, vol II, American Physiological Society, Bethesda, pp 877–946

Brown TG (1911) The intrinsic factor in the act of progression in the mammal. Proc R Soc Ser B 84:308–319

Brown TG (1914) On the nature of the fundamental activity of the nervous centres; together with an analysis of the conditioning of rhythmic activity in progression and a theory of the evolution of function in the nervous system. J Phsyiol (Lond) 48:18–46

Bruckmoser P, Hepp-Reymond M-C, Wiesendanger M (1970a) Cortical influence on single neurons of the lateral reticular nucleus of the cat. Exp Neurol 26: 239–252

Bruckmoser P, Hepp-Reymond M-C, Wiesendanger M (1970b) Effects of peripheral, rubral, and fastigial stimulation on neurons of the lateral reticular nucleus of the cat. Exp Neurol 27:388–398

Bruggencate G ten, Burke R, Lundberg A, Udo M (1969) Interaction between the vestibulospinal tract, contralateral flexor reflex afferents and Ia afferents. Brain Res 14:529–532

Bruggencate G ten, Teichmann R, Weller E (1972a) Neuronal activity in the lateral vestibular nucleus of the cat. I. Patterns of postsynaptic potentials and discharges in Deiters' neurones evoked by stimulation of the spinal cord. Pflügers Archiv Gesamte Physiol 337:119–134

Bruggencate G ten, Teichmann R, Weller E (1972b) Neuronal activity in the lateral vestibular nucleus of the cat. II. EPSPs in Deiters' neurones mediated by fast conducting fibres of the spinal cord. Pflügers Archiv Gesamte Physiol 337: 135–146

Bruggencate G ten, Teichmann R, Weller E (1972c) Neuronal activity in the lateral vestibular nucleus of the cat. III. Inhibitory actions of cerebellar Purkinje cells evoked via mossy and climbing fibre afferents. Pflügers Archiv Gesamte Physiol 337:147–162

Budakova NN (1973) Stepping movements in the spinal cat due to DOPA administration (in Russian). Fiziol Zh SSSR 59:1190–1198

Burke R, Jankowska E, Bruggencate G ten (1970) A comparison of peripheral and rubrospinal synaptic input to slow and fast twitch motor units of triceps surae. J Physiol (Lond) 207:709–732

Burke R, Lundberg A, Weight F (1971) Spinal border cell origin of the ventral spinocerebellar tract. Exp Brain Res 12:283–294

Burton JE, Onoda N (1977) Interpositus neuron discharge in relation to a voluntary movement. Brain Res 121:167–172

Burton JE, Onoda N (1978) Dependence of the activity of interpositus and red nucleus neurons on sensory input data generated by movement. Brain Res 152:41–64

Cabelguen J-M (1981) Static and dynamic fusimotor controls in various hindlimb muscles during locomotor activity in the decorticate cat. Brain Res 213:83–97

Clendenin M, Ekerot C-F, Oscarsson O, Rosén I (1974a) The lateral reticular nucleus in the cat. I. Mossy fibre distribution in cerebellar cortex. Exp Brain Res 21: 473–486

Clendenin M, Ekerot C-F, Oscarsson O, Rosén I (1974b) The lateral reticular nucleus in the cat. II. Organization of component activated from bilateral ventral flexor reflex tract (bVFRT). Exp Brain Res 21:487–500

Clendenin M, Ekerot C-F, Oscarsson O (1974c). The lateral reticular nucleus in the cat. III. Organization of component activated from ipsilateral forelimb tract. Exp Brain Res 21:501–513

Cody FWJ, Moore RB, Richardson HC (1981) Patterns of activity evoked in cerebellar interpositus nuclear neurones by natural somatosensory stimuli in awake cats. J Physiol (Lond) 317:1–20

Cohen MI, Feldman JL (1977) Models of respiratory phaseswitching. Fed Proc 36:2367–2374

Cooper S, Sherrington CS (1940) Gower's tract and spinal border cells. Brain 63: 123–134

Corvaja N, Grofova I, Pompeiano O, Walberg F (1977) The lateral reticular nucleus in the cat. I. An experimental anatomical study of its spinal and supraspinal afferent connections. Neuroscience 2:537–553

Courville J (1966) Somatotopical organization of the projection from the nucleus interpositus anterior of the cerebellum to the red nucleus. An experimental study in the cat with silver impregnation methods. Exp Brain Res 2:191–215

Courville J (1968) Connections of the red nucleus with the cerebellum and certain caudal brain stem structures. A review with functional considerations. Rev Biol 27:127–144

Courville J, Diakiw N (1976) Cerebellar corticonuclear projection in the cat. The vermis of the anterior and posterior lobes. Brain Res 110:1–20

Crichlow EC (1970) Contralateral projection of some cells of the lateral reticular nucleus. Canad J Physiol Pharmacol 48:569–572

Crichlow EC, Kennedy TT (1967) Functional characteristics of neurons in the lateral reticular nucleus with reference to localized cerebellar potentials. Exp Neurol 18:141–153

Crill WE, Kennedy TT (1967) Inferior olive of the cat: intracellular recording. Science 157:716–718

Curtis DR, Eccles J, Lundberg A (1958) Intracellular recording from cells in Clarke's column. Acta Physiol Scand 43:303–314

Deliagina TG (1977) The pathway of the scratch reflex in the cat (in Russian). Neirofiziologya 9:619–621

Deliagina TG, Feldman AG, Gelfand IM, Orlovsky GN (1975) On the role of central program and afferent inflow in the control of scratching movements in the cat. Brain Res 100:297–313

Deliagina TG, Orlovsky GN, Perret C (1981) Efferent activity during fictitious scratch reflex in the cat. J Neurophysiol 45:595–604

Deliagina TG, Orlovsky GN, Pavlova GA (1983) The capacity for generation of rhythmic oscillations is distributed in the lumbosacral spinal cord of the cat. Exp Brain Res 53:81–90

Dellow PG, Lund JP (1971) Evidence for central timing of rhythmical mastication. J Physiol (Lond) 215:1–13

Dietrichs E, Walberg F (1979) The cerebellar projection from the lateral reticular nucleus as studied with retrograde transport of horseradish peroxidase. Anat Embryol (Berl) 155:273–290

Domer FR, Feldberg W (1960) Scratching movements and facilitation of the scratch reflex produced by tubocurarine in cats. J Physiol (Lond) 153:35–51

Dow RS, Moruzzi G (1958) The physiology and pathology of the cerebellum. University Minnesota Press, Minneapolis

Eager RP (1963) Efferent cortico-nuclear pathways in the cerebellum of the cat. J Comp Neurol 120:81–103

Eager RP (1966) Patterns and mode of termination of cerebellar cortico-nuclear pathways in the monkey (Macaca mulatta). J Comp Neurol 126:551–566

Eager RP (1968) Some fine structural features of the neural elements composing the cerebellar nuclei in the cat. J Comp Neurol 132:235–241

Eccles JC (1970) The topography of the mossy and climbing fiber inputs to the anterior lobe of the cerebellum. In: Fields WS, Willis WD (eds) The cerebellum in health and disease, Warren H, Green, St. Louis, pp 231–266

Eccles JC (1973) The cerebellum as a computer: patterns in space and time. J Physiol (Lond) 229:1–32

Eccles JC (1977) An instruction-selection theory of learning in the cerebellar cortex. Brain Res 127:327–352

Eccles JC, Hubbart JI, Oscarsson O (1961a) Intracellular recording from cells of the ventral spinocerebellar tract. J Physiol (Lond) 158:486–516

Eccles JC, Oscarsson O, Willis W (1961b) Synaptic action of group I and II afferent fibres of muscle on the cells of the dorsal spino-cerebellar tract. J Physiol (Lond) 158:517–547

Eccles JC, Schmidt RF, Willis WD (1963) Inhibition of discharges into the dorsal and ventral spino-cerebellar tracts. J Neurophysiol 26:635–645

Eccles JC, Llinás R, Sasaki K (1966a) The inhibitory interneurones within the cerebellar cortex. Exp Brain Res 1:1–16

Eccles JC Llinás R, Sasaki K (1966b) Parallel fibre stimulation and the responses induced thereby in the Purkinje cells of the cerebellum. Exp Brain Res 1:17–39

Eccles JC, Llinás R, Sasaki K (1966c) The mossy fibre-granule cell relay in the cerebellum and its inhibition by Golgi cells. Exp Brain Res 1:82–101

Eccles JC, Llinás R, Sasaki K (1966d) The excitatory synaptic action of climbing fibres on the Purkinje cells of the cerebellum. J Physiol (Lond) 182:268–296

Eccles JC, Ito M, Szentágothai J (1967) The cerebellum as a neuronal machine. Springer, Berlin Heidelberg New York

Eccles JC, Provini L, Strata P, Tabořiková H (1968) Topographical investigations on the climbing fiber inputs from forelimb and hindlimb afferents to the cerebellar anterior lobe. Exp Brain Res 6:195–215

Eccles JC, Faber DS, Murphy JT, Sabah NH, Táboříková H (1971a) Afferent volleys in limb nerves influencing impulse discharges in cerebellar cortex. II. In Purkyně cells. Exp Brain Res 13:36–53

Eccles JC, Faber DS, Murphy JT, Sabah NH, Táboříková H (1971b) Investigations on integration of mossy fiber inputs to Purkyně cells in the anterior lobe. Exp Brain Res 13:54–77

Eccles JC, Rantucci T, Rosén I, Scheid P, Taboříková H (1974a) Somatotopic studies on cerebellar interpositus neurons. J Neurophysiol 37:1449–1459

Eccles JC, Rantucci T, Sabah NH, Táboříková H (1974b) Somatotopic studies on cerebellar fastigial cells. Exp Brain Res 19:100–118

Eccles JC, Rosén I, Scheid P, Táboříková H (1974c) Temporal patterns of responses of interpositus neurons to peripheral afferent stimulation. J Neurophysiol 37:1424–1437

Eccles JC, Rosén I, Scheid P, Táboříková H (1974d) Patterns of convergence onto interpositus neurons from peripheral afferents, J Neurophysiol 37:1438–1448

Eccles JC, Sabah NH, Táboříková H (1974e) Excitatory and inhibitory responses of neurones of the cerebellar fastigial nucleus. Exp Brain Res 19:61–77

Eccles JC, Sabah NH, Táboříková H (1974f) The pathways responsible for excitation and inhibition of fastigial neurones. Exp Brain Res 19:78–99

Eccles JC, Nicoll RA, Schwarz DWF, Táboříková H, Willey TJ (1975a) Reticulo-spinal neurons with and without monosynaptic inputs from cerebellar nuclei. J Neurophysiol 38:513–530

Eccles JC, Rantucci T, Scheid P, Táboříková H (1975b) Somatotopic studies on red nucleus: spinal projection level and respective receptive fields. J Neurophysiol 38:965–980

Eccles JC, Scheid P, Táboříková H (1975c) Responses of red nucleus neurons to antidromic and synaptic activation. J Neurophysiol 38:947–964

Eccles JC, Nicoll RA, Rantucci T, Táboříková H, Willey TJ (1976) Topographic studies on medial reticular nucleus. J Neurophysiol 39:109–118

Eide E, Fedina L, Jansen J, Lundberg A, Vyklicky L (1969a) Properties of Clarke's column neurones. Acta Physiol Scand 77:125–144

Eide E, Fedina L, Jansen J, Lundberg A, Vyklicky L (1969b) Unitary components in the activation of Clarke's column neurones. Acta Physiol Scand 77:145–158

Eidelberg E, Story JL, Walden JG, Meyer BL (1981) Anatomical correlates of return of locomotor function after partial spinal cord lesions in cats. Exp Brain Res 42:81–88

Ekerot C-F, Larson B, Oscarsson O (1979) Information carried by the spinocerebellar paths. In: Granit R, Pompeiano O (eds) Progress in Brain Research, vol 50, Reflex control of posture and movement. Elsevier, Amsterdam, pp 79–90

Engberg I, Lundberg A (1969) An electromyographic analysis of muscular activity in the hindlimb of the cat during unrestrained locomotion. Acta Physiol Scand 75:614–630

Euler C v (1977) The functional organization of the respiratory phase switching mechanisms. Fed Proc 36:2375–2380

Euler C v (1983) On the origin and pattern control of breathing rhythmicity in mammals. In: Roberts A, Roberts B (eds) Neural origin of rhythmic movements. Cambridge University Press, pp 469–485

Evarts EV (1966) Methods for recording activity of individual neurones in moving animals. In: Rushmer RF (ed) Methods in medical research, vol II, Year Book Medical Publishers, Chicago, pp 241–250

Evarts EV (1968) A technique for recording activity of subcortical neurons in moving animals. Electroencephalogr Clin Neurophysiol 24:83–86

Evarts EV, Thach WT (1969) Motor mechanisms of the CNS: cerebrocerebellar interrelations. Annu Rev Physiol 31:451–498

Fanardjian VV (1975) On neuronal organization of cerebellar efferent systems (in Russian). Nauka, Leningrad

Fanardjian VV, Manvelian IA (1976) Peculiarities of the cutaneous sensitivity representation in the red nucleus of the cat (in Russian). Fiziol Zh SSSR 62: 499–509

Fanardjian VV, Sarkisjian DS (1969) Intracellular investigation of antidromic and synaptic activation of red nuclear neurons in the cat (in Russian). Fiziol Zh SSSR 55:121–131

Fanardjian VV, Sarkissian VA (1980) Spatial organization of the cerebellar cortico-vestibular projection in the cat. Neuroscience 5:551–558

Feldberg W, Fleischhauer K (1960) Scratching movements evoked by drugs applied to the upper cervical cord. J Physiol (Lond) 151:502–517

Feldman AG (1979) Central and reflex mechanisms of the movement control (in Russian). Nauka, Moscow

Feldman AG, Orlovsky GN (1975) Activity of interneurons mediating reciprocal Ia inhibition during locomotion. Brain Res 84:181–194

Feldman AG, Orlovsky GN, Perret C (1977) Activity of muscle spindle afferents during scratching in the cat. Brain Res 129:192–196

Flood S, Jansen J (1961) On the cerebellar nuclei in the cat. Acta Anat (Basel) 46:52–72

Flumerfelt BA (1978) Organization of the mammalian red nucleus and its interconnections with the cerebellum. Experientia (Basel) 34:1178–1180

Forssberg H, Grillner S (1973) The locomotion of the acute spinal cat injected with Clonidine i.v. Brain Res 50:184–186

Forssberg H, Grillner S, Rossignol S (1975) Phase dependent reflex reversal during walking in chronic spinal cord. Brain Res 85:103–107

Forssberg H, Grillner S, Halbertsma J (1980a) The locomotion of the low spinal cat. I. Coordination within a hindlimb. Acta Physiol Scand 108:269–281

Forssberg H, Grillner S, Halbertsma J, Rossignol S (1980b) The locomotion of the low spinal cat. II. Interlimb coordination. Acta Physiol Scand 108:283–295

Fox CA (1962) The structure of the cerebellar cortex. In: Crosby EC, Humphrey T, Lauer EW (eds) Correlative anatomy of the nervous system. Macmillan, New York, pp 193–198

Fox CA, Snider RS (eds) (1967) Progress in brain research, vol 25, The Cerebellum. Elsevier, Amsterdam

Freusberg A (1874) Reflexbewegungen beim Hunde. Pflügers Archiv Gesamte Physiol 9:358–391

Fu T-C, Jankowska E, Tanaka R (1977) Effects of volleys in cortico-spinal tract fibres on ventral spino-cerebellar tract cells in the cat. Acta Physiol Scand, 100:1–13

Fukushima K, Peterson BW, Wilson VJ (1979) Vestibulospinal, reticulospinal and interstitiospinal pathways in the cat: In: Granit R, Pompeiano O (eds) Progress in brain research, vol 50, Reflex control of posture and movement. Elsevier Amsterdam, pp 121–136

Gambarian PP, Orlovsky GN, Protopopova TG, Severin FV, Shik ML (1971) Muscular activity at different kinds of cat locomotion, and adjusting of the organs of movements in the family Felidae (in Russian). Proc Zool Inst Acad Sci SSSR 48:220–239

Gelfand IM, Zetlin ML (1966) On the mathematical modelling of mechanisms of the central nervous system. In: Gelfand IM et al. (eds) Models of the structural-functional organization of certain biological systems (in Russian) (English translation by MIT Press, Massachusetts 1971), Nauka, Moscow

Gernandt BE, Thulin C-A (1953) Vestibular mechanisms of facilitation and inhibition of cord reflexes. Am J Physiol 172:653–660

Gernandt BE, Thulin C-A (1955) Reciprocal effects upon spinal motoneurones from stimulation of bulbar reticular formation. J Neurophysiol 18:113–129

Gesell R, Bricker J, Magee S (1936) Structural and functional organization of the central mechanism controlling breathing. Am J Physiol 117:423–452

Ghez C (1975) Input-output relations of the red nucleus in the cat. Brain Res 98: 93–108

Ghez C, Kubota K (1977) Activity of red nucleus neurons associated with a skilled forelimb movement in the cat. Brain Res 131:383–388

Ghez C, Vicario D (1978) Discharge of red nucleus neurons during voluntary muscle contraction: activity patterns and correlations with isometric force. J Physiol (Paris) 74:283–285

Gilbert PFC, Thach WT (1977) Purkinje cell activity during motor learning. Brain Res 128:309–328

Giuffrida R, Volsi GL, Panto MR, Perciavalle V, Sapienza S, Urbano A (1980) Single muscle organization of interpositorubral projections. Exp Brain Res 39:261–267

Giuffrida R, Volsi GL, Perciavalle V, Santangelo F, Urbano A (1981) Influences of precerebellar systems triggering movement on single cells of the interpositus nucleus of the cat. Neuroscience 6:1625–1631

Granit R (1970) The basis of motor control. Academic, London

Grant G (1962) Spinal course and somatotopically localized termination of the spinocerebellar tracts. An experimental study in the cat. Acta Physiol Scand 56, Suppl 193:1–45

Grant G, Oscarsson O, Rosén I (1966) Functional organization of the spino-reticulo-cerebellar path with identification of its spinal component. Exp Brain Res 1:306–319

Grillner S (1969) Supraspinal and segmental control of static and dynamic gamma-motoneurons in the cat. Acta Physiol Scand 77, Suppl 327:1–34

Grillner S (1972) The role of muscle stiffness in meeting the changing postural and locomotor requirements for force development by the ankle extensors. Acta Physiol Scand, 86:92–108

Grillner S (1973) Locomotion in the spinal cat. In: Stein RB, Pearson KY, Smith RS, Redford JB (eds), Control of posture and locomotion, Plenum, New York, pp 515–535

Grillner S (1975) Locomotion in vertebrates: Central mechanisms and reflex interaction. Physiol Rev 55:247–304

Grillner S (1981) Control of locomotion in bipeds, tetrapods, and fish. In: Brooks VB (ed) Handbook of physiology, the nervous system, v II, American Physiological Society, Bethesda, pp 1179–1236

Grillner S, Hongo T (1972) Vestibulospinal effects on motoneurones and interneurones in the lumbosacral cord. In: Brodal A, Pompeiano O (eds) Progress in brain research, vol 37, Basic aspects of central vestibular mechanisms, Elsevier, Amsterdam, pp 243–262

Grillner S, Lund S (1968) The origin of a descending pathway with monosynaptic action on flexor motoneurones. Acta Physiol Scand, 74:274–284

Grillner S, Rossignol S (1978) On the initiation of the swing phase of locomotion in chronic spinal cats. Brain Res 146:269–277

Grillner S, Zangger P (1979) On the central generation of locomotion in the low spinal cat. Exp Brain Res 34:241–261

Grillner S, Hongo T, Lund S (1966a) Interaction between the inhibitory pathways from the Deiters' nucleus and Ia afferents to flexor motoneurones. Acta Physiol Scand, 68, Suppl 277:61

Grillner S, Hongo T, Lund S (1966b) Monosynaptic excitation of spinal $\gamma$-motoneurones from the brain stem. Experientia (Basel) 22:691

Grillner S, Hongo T, Lund S (1968a) The origin of descending fibres monosynaptically activating spinoreticular neurones. Brain Res 10:259–262

Grillner S, Hongo T, Lund S (1968b) Reciprocal effects between two descending bulbospinal systems with monosynaptic connections to spinal motoneurones. Brain Res 10:477–480

Grillner S, Hongo T, Lund S (1969) Descending monosynaptic and reflex control of γ-mononeurones. Acta Physiol Scand, 75:592–613

Grillner S, Hongo T, Lund S (1970) The vestibulospinal tract. Effects on alpha-motoneurones in the lumbosacral spinal cord in the cat. Exp Brain Res 10: 94–120  -

Grillner S, Hongo T, Lund S (1971) Convergent effects on alpha-motoneurones from the vestibulospinal tract and a pathway descending in the medial longitudinal fasciculus. Exp Brain Res 12:457–479

Grimm RJ, Rushmer DS (1974) The activity of dentate neurons during an arm movement sequence. In: Motor aspects of behaviour and programmed nervous activities. Brain Res 71:309–326

Gurfinkel VS, Shik ML (1973) The control of posture and locomotion. In: Gydikov AA, Tankov UT, Kosarov DS (eds) Motor control, Plenum, New York, pp 217–234

Gurfinkel VS, Kotz YaM, Shik ML (1965) Control of the human posture (in Russian). Nauka, Moscow

Gurfinkel VS, Kostyuk PG, Shik ML (1973) On some possible modes of descending control of the spinal cord activity in connection with the problem of motor control. In: Proc Symp 4th Int Biophys Congr, Moscow

Gustafsson B, Lindström S (1973) Recurrent control from motor axon collaterals of Ia inhibitory pathways to ventral spinocerebellar tract neurons. Acta Physiol Scand, 89:457–481

Hampson JL, Harrison CR, Woolsey CN (1952) Cerebro-cerebellar projections and the somatotopic localization of motor function in the cerebellum. Assoc Res Nerv Ment Dis, Proc, 30:299–316

Hare WK, Magoun HW, Ranson SW (1937) Localization within the cerebellum of reactions to faradic cerebellar stimulation. J Comp Neurol 67:145–182

Hart BL (1971) Facilitation by strychnine of reflex walking in spinal dogs. In: Physiology and behaviour, vol 6, Pergamon, New York, pp 627–628

Harvey RJ, Porter R, Rawson JA (1977) The natural discharges of Purkinje cells in paravermal regions of lobules V and VI of the monkey's cerebellum. J Physiol (Lond) 271:515–536

Harvey RJ, Porter R, Rawson JA (1979) Discharges of intracerebellar nuclear cells in monkeys. J Physiol (Lond) 297:559–580

Hernandez-Mesa N, Bureš J (1978) Skilled forelimb movements and unit activity of cerebellar cortex and dentate nucleus in rats. Physiol Bohemoslov 27:199–208

Hikosaka M, Maeda M, Nakao S, Shimazu H, Shinoda Y (1977) Presynaptic impulses in the abducens nucleus and their relation to postsynaptic potentials in motoneurons during vestibular nystagmus. Exp Brain Res 27:355–376

Hökfelt T, Fuxe K (1969) Cerebellar monoamine nerve terminals, a new type of afferent fibers to the cortex cerebelli. Exp Brain Res 9:63–72

Holmes G (1939) The cerebellum of man. Brain, 62:1–30

Holmqvist B, Lundberg A, Oscarsson O (1956) Functional organization of the dorsal spino-cerebellar tract in the cat. V. Further experiments on convergence of excitatory and inhibitory actions. Acta Physiol Scand 38:76–90

Holmqvist B, Lundberg A, Oscarsson O (1960) Supraspinal inhibitory control of transmission to three ascending spinal pathways influenced by the flexion reflex afferents. Arch Ital Biol 98:60–80

Hongo T, Okada Y (1967) Cortically evoked pre- and postsynaptic inhibition of impulse transmission to the dorsal spinocerebellar tract. Exp Brain Res 3: 163–177

Hongo T, Okada Y, Sato M (1967) Corticofugal influences on transmission to the dorsal spinocerebellar tract from hindlimb primary afferents. Exp Brain Res 3:135–149

Hongo T, Jankowska E, Lundberg A (1969a) The rubrospinal tract. I. Effects on alpha-motoneurones innervating hindlimb muscles in cats. Exp Brain Res 7: 344–364

Hongo T, Jankowska E, Lundberg A (1969b) The rubrospinal tract. II. Facilitation of interneuronal transmission in reflex paths to motoneurons. Exp Brain Res 7:365–391

Hongo T, Kudo N, Tanaka R (1975) The vestibulospinal tract: crossed and uncrossed effects on hindlimb motoneurons in the cat. Exp Brain Res 24:37–55

Hongo T, Jankowska E, Ohno T, Sasaki S, Yamashita M, Yoshida K (1983a) Inhibition of dorsal spinocerebellar tract cells by interneurons in upper and lower lumber segments in the cat. J Physiol (Lond) 342:145–159

Hongo T, Jankowska E, Ohno T, Sasaki S, Yamashita M, Yoshida K (1983b) The same interneurons mediate inhibition of dorsal spinocerebellar tract cells and lumbar motoneurones in the cat. J Physiol (Lond) 342:161–180

Hubbard JI, Oscarsson O (1962) Localization of the cell bodies of the ventral spinocerebellar tract in lumber segments of the cat. J Comp Neurol 118:199–204

Illert M, Lundberg A, Padel Y, Tanaka R (1978) Integration in descending motor pathways controlling the forelimb in the cat. 5. Properties of and monosynaptic excitatory convergence on C3–C4 propriospinal neurons. Exp Brain Res 33: 101–130

Ito M (1965) Origin of cerebellar inhibition of Deiters' and intracerebellar nuclei. In: Curtis DR, McIntyre AK (eds) Studies in physiology. Presented to JC Eccles, Springer, Berlin Heidelberg New York, pp 100–106

Ito M (1967) Neuronal circuitry in the cerebellar efferent system. In: Yahr MD, Purpura DP (eds) Neurophysiological basis of normal and abnormal motor activities. Raven, New York, pp 119–140

Ito M (1982) Experimental verification of Marr-Albus' plasticity assumption for the cerebellum. Acta Biol Acad Sci Hung 33:189–199

Ito M, Kano M (1982) Long-lasting depression of parallel fiber-Purkinje cell transmission induced by conjunctive stimulation of parallel fibers and climbing fibers in the cerebellar cortex. Neurosci Lett 33:253–258

Ito M, Yoshida M (1964) The cerebellar-evoked monosynaptic inhibition of Deiters' neurones. Experientia (Basel) 20:515–516

Ito M , Yoshida M (1966) The origin of cerebellar-induced inhibition of Deiters' neurones. I. Monosynaptic initiation of the inhibitory postsynaptic potentials. Exp Brain Res 2:330–349

Ito M, Hongo T, Yoshida M, Okada Y, Obata K (1964a) Antidromic and transsynaptic activation of Deiters' neurones induced from the spinal cord. Jpn J Physiol 14:638–658

Ito M, Yoshida M, Obata K (1964b) Monosynaptic inhibition of the intracerebellar nuclei induced from the cerebellar cortex. Experientia (Basel) 20:575–576

Ito M, Obata K, Ochi R (1966) The origin of cerebellar-induced inhibition of Deiters' neurones. II. Temporal correlation between the trans-synaptic activation of Purkinje cells and the inhibition of Deiters' neurones. Exp Brain Res 2:350–364

Ito M, Kawai N, Udo M (1968a) The origin of cerebellar-induced inhibition of Deiters' neurones. III. Distribution of the inhibitory zone. Exp Brain Res 4: 310–320

Ito M, Kawai N, Udo M, Sato N (1968b) Cerebellar-evoked disinhibition in dorsal Deiters' neurones. Exp Brain Res 6:247–264

Ito M, Kawai N, Udo M, Mano N (1969) Axon reflex activation of Deiters' neurones from the cerebellar cortex through collaterals of the cerebellar afferents. Exp Brain Res 8:249–268

Ito M, Udo M, Mano N (1970a) Long inhibitory and excitatory pathways converging onto cat reticular and Deiters' neurones and their relevance to reticulofugal axons. J Neurophysiol 33:210–226

Ito M, Udo M, Mano N, Kawai N (1970b) Synaptic action of the fastigiobulbar impulses upon neurones in the medullary reticular formation and vestibular nuclei. Exp Brain Res 11:29–47

Ito M, Yoshida M, Obata K, Kawai N, Udo M (1970c) Inhibitory control of intra-
    cerebellar nuclei by the Purkinje cell axons. Exp Brain Res 10:64–80
Jankowska E, Lund S, Lundberg A, Pompeiano O (1968) Inhibitory effects evoked
    through ventral reticulospinal pathways. Arch Ital Biol 106:124–140
Jansen J, Brodal A (1940) Experimental studies on the intrinsic fibres of the cere-
    bellum. II. The cortico-nuclear projection. J Comp Neurol 73:267–321
Jansen J, Brodal A (1954) Aspects of cerebellar anatomy. Johan Grundt Tanum, Oslo
Jansen JKS, Rudjord T (1965) Dorsal spinocerebellar tract: Response pattern of
    nerve fibers to muscle stretch. Science 149:1109–1111
Jansen JKS, Wallöe L (1970) Transmission of signals from muscle stretch receptors
    to the dorsal spino-cerebellar tract. In: Fields WS, Willis WD (eds) The cerebel-
    lum in health and disease. Warren H Green, St. Louis, pp 143–171
Jansen JKS, Nicolaysen K, Rudjord T (1966) The discharge pattern of neurons of
    the dorsal spinocerebellar tract activated by static extension of the primary
    endings of muscle spindles. J Neurophysiol 29:1061–1086
Jansen JKS, Nicolaysen K, Wallöe L (1967a) On the inhibition of transmission to
    the dorsal spinocerebellar tract by stretch of various ankle muscles of the cat.
    Acta Physiol Scand 70:362–368
Jansen JKS, Poppele RE, Terzuolo CA (1967b) Transmission of proprioceptive
    information via the dorsal spinocerebellar tract. Brain Res 6:382–384
Jordan LM, Pratt CA, Menzies E (1979) Locomotion evoked by brain stem stimula-
    tion: occurence without phasic segmental afferent input. Brain Res 177:204–207
Kandel ER (1976) The cellular basis of behaviour. Freeman, San Francisco
Kazennikov OV, Selionov VA, Shik ML, Yakovleva GV (1979) Neurons of upper
    cervical segments responding to stimulation of the bulbar "locomotor strip"
    (in Russian). Neirofiziologya 11:245–253
Kazennikow OV, Shik ML, Yakovleva GV (1980) Two pathways for the brain stem
    "locomotor influence" on the spinal cord (in Russian). Fiziol Zh SSSR 66:
    1260–1263
Kawaguchi S, Ohno T (1974) Responses of interpositus neurones to inputs from
    muscle receptors. Exp Brain Res 21:375–386
Kennedy D, Davis WJ (1977) Organization of invertebrate motor systems. In:
    Kandel ER (ed) Handbook of physiology. Sec 1, The nervous systems, vol 1,
    part 2, American Physiological Society, Bethesda, pp 1023–1087
Kitai ST, Tábořiková H, Tsukahara N, Eccles JC (1969) The distribution to the
    cerebellar anterior lobe of the climbing and mossy fiber inputs from the plantar
    and palmar cutaneous afferents. Exp Brain Res 7:1–10
Kitai ST, DeFrance JF, Hatada K, Kennedy DT (1974) Electrophysiological pro-
    perties of lateral reticular nucleus cells: II. Synaptic activation. Exp Brain Res
    21:419–432
Kitai ST, McCrea RA, Preston RJ, Bishop GA (1977) Electrophysiological and
    horseradish peroxidase studies of precerebellar afferents to the nucleus inter-
    positus anterior. I. Climbing fiber system. Brain Res 122:197–214
Kostyuk PG (1969) On the functions of dorsal spino-cerebellar tract in cat. In:
    Llinás R (ed) Neurobiology of cerebellar evolution and development. American
    Medical Association, Chicago, pp 539–548
Kostyuk PG (1973) Structure and function of descending systems of the spinal
    cord (in Russian). Nauka, Leningrad
Kostyuk PG, Pilyavsky AI (1969) Synaptic processes in spinal interneurones during
    rubrospinal influences (in Russian). Neirofiziologya 1:158–166
Kubota S, Poppele RE (1977) Evidence for control of DSCT activity by the brain
    stem reticular formation. Brain Res 129:361–365
Kuno M, Miyahara JT (1968) Factors responsible for multiple discharge of neurones
    in Clarke's column. Neurophysiol 31:624–636

Künzle H (1973) The topographic organization of spinal afferents to the lateral reticular nucleus of the cat. J Comp Neurol 149:103—116

Kuypers HGJM, Fleming WR, Farinholt JW (1962) Subcorticospinal projections in the rhesus monkey. J Comp Neurol 118:107—137

Ladpli R, Brodal A (1968) Experimental studies of comissural and reticular formation projections from the vestibular nuclei in the cat. Brain Res 8:65—96

Laporte Y, Lundberg A, Oscarsson O (1956a) Functional organization of dorsal spino-cerebellar tract in the cat. I. Recording of mass discharge in dissected Flechsig's fasciculus. Acta Physiol Scand 36:175—187

Laporte Y, Lundberg A, Oscarsson O (1956b) Functional organization of the dorsal spino-cerebellar tract in the cat. II. Single fibre recording in Flechsig's fasciculus. Acta Physiol Scand 36:188—203

Larsell O (1937) The cerebellum. A review and interpretation. Arch Neurol Psychiatry (Chicago) 58:580—607

Larsell O (1967) The comparative anatomy and histology of the cerebellum from mixinoids through birds. University of Minnesota Press, Minneapolis

Larsell O, Jansen J (1972) The comparative anatomy and histology of the cerebellum. The human cerebellum, cerebellar connections and cerebellar cortex. University of Minnesota Press, Minneapolis

Larsen KD, Yumiya H (1980) The red nucleus of the monkey. Topographic localization of somatosensory input and motor output. Exp Brain Res 40:393—404

Larson B, Miller S, Oscarsson O (1969a) Termination and functional organization of the dorsolateral spino-olivocerebellar path.J Physiol (Lond) 203:611—640

Larson B, Miller S, Oscarsson O (1969b) A spinocerebellar climbing fibre path activated by the flexor reflex afferents from all four limbs. J Physiol (Lond) 203:641—649

Latham A, Paul DH (1971) Effects of sodium thiopentone on cerebellar neurone activity. Brain Res 25:212—215

Leicht R, Rowe MJ, Schmidt RF (1977) Mossy and climbing fiber inputs from cutaneous mechanoreceptors to cerebellar Purkyně cells in unanesthetized cats. Exp Brain Res 27:459—477

Lennard PR, Getting PA, Hume RI (1980) Central pattern generator mediating swimming in Tritonia. II. Initiation, maintenance, and termination. J Neurophysiol 44:165—173

Lindsley DB (1952) Brain stem influences on spinal motor activity. Res Publ Assoc Nerv Ment Dis 30:174—195

Lindström S (1973) Recurrent control from motor axon collaterals of Ia inhibitory pathways in the spinal cord of the cat. Acta Physiol Scand. Suppl 392:1—43

Lindström S, Schomburg ED (1973) Recurrent inhibition from motor axon collaterals of ventral spinocerebellar tract neurons. Acta Physiol Scand 88:505—515

Lindström S, Schomburg ED (1974) Group I inhibition in Ib excited ventral spinocerebellar tract neurons.Acta Physiol Scand, 90:166—185

Lindström S, Takata M (1972) Monosynaptic excitation of dorsal spinocerebellar tract neurons from low threshold joint afferents. Acta Physiol Scand 84:430—432

Llinás R (1964) Mechanisms of supraspinal actions upon spinal cord activities. Differences between reticular and cerebellar inhibitory actions upon alpha extensor motoneurons. J Neurophysiol 27:1117—1126

Llinás R (ed) (1969) Neurobiology of cerebellar evolution and development. American Medical Association, Chicago

Llinás R (1974) Motor aspects of cerebellar control. Physiologist 17:19—46

Llinás R, Terzuolo CA (1964) Mechanisms of supraspinal actions upon spinal cord activities. Reticular inhibitory mechanisms in alpha extensor motoneurones. J Neurophysiol 27:579—590

Llinás R, Terzuolo CA (1965) Mechanisms of supraspinal actions upon spinal cord activities. Reticular inhibitory mechanisms upon flexor motoneurones. J Neurophysiol 28 413—422

Lloyd DPC, McIntyre AK (1950) Dorsal column conduction of group I muscle afferent impulses and their relay through Clarke's column. J Neurophysiol 13:39–54

Loeb GE, Duysens J (1979) Activity patterns in individual hindlimb primary and secondary muscle spindle afferents during normal movements in unrestrained cats. J Neurophysiol 42:420–440

Lorente de No R (1933) Vestibulo-ocular reflex arc. Arch Neurol Psychiatr 30: 245–291

Luciani L (1915) Muscular and nervous system. In: Human physiology, vol 3. McMillan, London

Lund JP (1976) Evidenc for a central neural pattern generator regulating the chewing cycle. In: Anderson DJ, Matthews DB (eds) Mastication, Wright, Bristol, pp 204–212

Lund S, Pompeiano O (1965) Descending pathways with monosynaptic action on motoneurones. Experientia (Basel) 21:602–603

Lund S, Pompeiano O (1968) Monosynaptic excitation of α-motoneurones from supraspinal structures in the cat. Acta Physiol Scand 73:1–21

Lundberg A (1959) Integrative significance of patterns of connections made by muscle afferents in the spinal cord. In: Symp Lect XXI Int Congr Physiol Sci, Buenos Aires, pp 100–105

Lundberg A (1964) Ascending spinal hindlimb pathways in the cat. In: Eccles JC, Schade JP (eds) Progress in brain research, vol 12, Elsevier, Amsterdam, pp 135–163

Lundberg A (1966) Integration in the reflex pathway. In: Granit R (ed) Nobel symposium. I. Muscular afferents and motor control, Almqvist and Wiksell, Stockholm, pp 275–305

Lundberg A (1969) Reflex control of stepping. In: The Nansen memorial lecture, V, Universitetsforlaget, Oslo, pp 1–42

Lundberg A (1971) Function of the ventral spinocerebellar tract. A new hypothesis. Exp Brain Res 12:317–330

Lundberg A (1981) Half-centres revisited. In: Symp lect of XXVIII Int Congr Physiol Sci, Akadémiai Kiadó, Budapest, pp 155–167

Lundberg A, Norrsell U, Voorhoeve P (1963) Effects from the sensorimotor cortex on ascending spinal pathways. Acta Physiol Scand 59:462–473

Lundberg A, Oscarsson O (1956) Functional organization of the dorsal spino-cerebellar tract in the cat. IV. Synaptic connections of afferents from Golgi tendon organs and muscle spindles. Acta Physiol Scand 38:53–75

Lundberg A, Oscarsson O (1960) Functional organization of the dorsal spino-cerebellar tract in the cat. VII. Identification of units by antidromic activation from the cerebellar cortex with recognition of five functional subdivisions. Acta Physiol Scand 50:356–374

Lundberg A, Oscarsson O (1962a) Functional organization of the ventral spino-cerebellar tract in the cat. IV. Identification of units by antidromic activation from the cerebellar cortex. Acta Physiol Scand 54:252–269

Lundberg A, Oscarsson O (1962b) Two ascending spinal pathways in the ventral part of the cord. Acta Physiol Scand 54:270–286

Lundberg A, Weight F (1971) Functional organization of connexions to the ventral spino-cerebellar tract. Exp Brain Res 12:295–316

Lundberg A, Winsbury G (1960) Functional organization of the dorsal spino-cerebellar tract in the cat. VI. Further experiments on excitation from tendon organ and muscle spindle afferents. Acta Physiol Scand 49:165–170

MacKay WA, Murphy JT (1974) Responses of interpositus neurons to passive muscle stretch. J Neurophysiol 37:1410–1423

Magni F, Oscarsson O (1961) Cerebral control of transmission to the ventral spino-cerebellar tract. Arch Ital Biol 99:369–396

Magni F, Willis WD (1963) Identification of reticular formation neurons by intra-
cellular recording. Arch Ital Biol 101:681–702
Magni F, Willis WD (1964a) Subcortical and peripheral control of brain stem reticular
neurons. Arch Ital Biol 102:434–448
Magni F, Willis WD (1964b) Afferent connections to reticulospinal neurons. In:
Eccles JC, Schade JP (eds) Progress in brain research, vol 12, Physiology of
spinal neurons. Elsevier, Amsterdam
Magnus R (1925) Animal posture. Proc R Soc Lond Ser B 98:339–353
Magoun HW, Rhines R (1946) An inhibitory mechanism in the bulbar reticular
formation. J Neurophysiol 9:165–171
Mann MD (1973) Clarke's column and the dorsal spino-cerebellar tract: a review.
Brain Behav Evol 7:34–83
Mano N (1974) Simple and complex spike activities of the cerebellar Purkinje cell
in relation to selective alternate movement in intact monkey. Brain Res 70:
381–393
Mano N (1979) Analysis of cerebellar Purkinje cell activity in relation to the direction,
position and velocity of wrist tracking movement. In: Massion J, Sasaki K (eds)
Developments in neurosciences, vol 6, Cerebro-cerebellar interactions. Elsevier,
Amsterdam, pp 163–183
Mano N, Yamamoto K (1975) Slow and rapid (quick or ballistic) tracking move-
ments and a cerebellar Purkinje cell activity. J Physiol Soc Jpn 37:383–384
Mano N, Yamamoto K (1980) Simple-spike activity of cerebellar Purkinje cells
related to visually guided wrist tracking movement in the monkey. J Neuro-
physiol 43:713–728
Marr D (1969) A theory of cerebellar cortex. J Physiol (Lond) 202:437–470
Massion J (1967) The mammalian red nucleus. Physiol Rev 47:383–436
Massion J, Sasaki K (1979) Cerebro-cerebellar interaction: solved and unsolved
problems. In: Massion J, Sasaki K (eds) Developments in neurosciences, vol 6,
Cerebro-cerebellar interactions. Elsevier, Amsterdam, pp 261–287
Matsushita M, Ikeda M (1970a) Olivary projections to the cerebellar nuclei in the
cat. Exp Brain Res 10:488–500
Matsushita M, Ikeda M (1970b) Spinal projections to the cerebellar nuclei in the cat.
Exp Brain Res 10:501–511
Matsushita M, Ikeda M (1976) Projections from the lateral reticular nucleus to the
cerebellar cortex and nuclei in the cat. Exp Brain Res 24:403–421
Matsushita M, Iwahori N (1971a) Structural organization of the fastigial nucleus.
I. Dendrites and axonal pathways. Brain Res 25:597–610
Matsushita M, Iwahori N (1971b) Structural organization of the interpositus and
dentate nuclei. Brain Res 35:17–36
Matthews PBS (1972) Mammalian muscle receptors and their central action. Arnold,
London
McCrea RA, Bishop GA, Kitai ST (1977) Electrophysiological and horseradish
peroxidase studies of precerebellar afferents to the nucleus interpositus anterior.
II. Mossy fiber system. Brain Res 122:215–228
McElligott JG (1976) Cerebellar neuronal fibring patterns in the intact and unrestrained
cat during walking. In: Herman RM, Grillner S, Stein PSG, Stuart DG (eds)
Neural control of locomotion. Plenum, New York, pp 781–784
McIntyre AK, Mark RF (1960) Synaptic linkage between afferent fibres of the
cat's hindlimb and ascending fibres in the dorsolateral funiculus. J Physiol
(Lond) 153:306–330
Mehler WR, Feferman ME, Nauta WJH (1960) Ascending axon degeneration follow-
ing anterolateral chordotomy. An experimental study in the monkey. Brain
83:718–750
Meyer-Lohmann J, Hore J, Brooks VB (1977) Cerebellar participation in generation
of prompt arm movements. J Neurophysiol 40:1038–1050

Miller AJ (1972) Significance of sensory inflow to the swallowing reflex. Brain Res 43:147–159

Miller S, Oscarsson O (1970) Termination and functional organization of spino-olivocerebellar paths. In: Fields WS, Willis WD (eds) The cerebellum in health and disease. Warren H Green, St Louis, pp 172–200

Miller S, Nezlina N, Oscarsson O (1969a) Projection and convergence patterns in climbing fibre paths to cerebellar anterior lobe activated from cerebral cortex and spinal cord. Brain Res 14:230–233

Miller S, Nezlina N, Oscarsson O (1969b) Climbing fibre projection to cerebellar anterior lobe activated from structures in midbrain and from spinal cord. Brain Res 14:234–236

Monakow C (1883) Experimenteller Beitrag zur Kenntnis des Corpus restiforme, des "äußeren Acusticuskerns" und deren Beziehungen zum Rückenmark. Arch Psychiatr Nervenkr 14:1–16

Morin F, Kennedy DT, Gardner E (1966) Spinal afferents to the lateral reticular nucleus. I. An histological study. J Comp Neurol 126:511–522

Moruzzi G (1954) The physiological properties of the brain stem. In: Delafresnay JF (ed) Brain mechanisms and consciousness. Blackwell, Oxford, pp 21–53

Murphy JT, Sabah NH (1971) Cerebellar Purkinje cell responses to afferent inputs. I. Climbing fiber activation. Brain Res 25:449–467

Murphy JT, MacKay WA, Johnson F (1973) Differences between cerebellar mossy and climbing fibre responses to natural stimulation of forelimb muscle proprioceptors. Brain Res 55:263–289

Nakamura Y, Mizuno N (1971) An electron microscopic study of the interposito-rubral connections of the cat and rabbit. Brain Res 35:283–286

Nakamura Y, Mizuno N, Konishi A (1978) A quantitative electron microscope study of cerebellar axon terminals on the magnocellular red nucleus neurons in the cat. Brain Res 147:17–27

Niemer WT, Magoun HW (1947) Reticulo-spinal tract influencing motor activity. J Comp Neurol 87:367–379

Nishioka S, Nakahama H (1973) Peripheral somatic activation of neurons in the cat red nucleus. J Neurophysiol 36:296–307

Nyberg-Hansen R (1964) Origin and termination of fibers from the vestibular nuclei descending in the medial longitudinal fasciculus. An experimental study with silver impregnation methods in the cat. J Comp Neurol 122:355–367

Nyberg-Hansen R (1965) Sites and mode of termination of reticulospinal fibers in the cat. An experimental study with silver impregnation methods. J Comp Neurol 124:71–100

Nyberg-Hansen R (1966) Functional organization of descending supraspinal fibre systems to the spinal cord. Anatomical observations and physiological correlations. Ergeb Anat Entwicklungsgesch 39:1–48

Nyberg-Hansen R, Brodal A (1964) Site and mode of termination of rubrospinal fibers in the cat. An experimental study with silver impregnation method. J Anat (Lond) 98:235–253

Nyberg-Hansen R, Mascitti TA (1964) Sites and mode of termination of fibers of the vestibulospinal tract in the cat. An experimental study with silver impregnation method. J Comp Neurol 122:369–388

O'Donovan MJ, Pinter MJ, Dum RP, Burke RE (1982) Actions of FDL and FHL muscles in intact cats: functional dissociation between anatomical synergists. J Neurophysiol 47:1126–1143

Olson L, Fuxe K (1971) On the projections from the locus coeruleus noradrenaline neurons: The cerebellar innervations. Brain Res 28:165–171

Orlovsky GN (1969) Spontaneous and induced locomotion of the thalamic cat. Biophysics 14:1154–1162

Orlovsky GN (1970a) Connexions of the reticulo-spinal neurons with the "locomotor regions" of the brain stem. Biophysics 15:178–186

Orlovsky GN (1970b) Work of the reticulo-spinal neurones during locomotion. Biophysics 15:761—771

Orlovsky GN (1970c) Influence of the cerebellum on the reticulo-spinal neurones during locomotion. Biophysics 15:928—936

Orlovsky GN (1972a) The effect of different descending systems on flexor and extensor activity during locomotion. Brain Res 40:359—371

Orlovsky GN (1972b) Activity of vestibulospinal neurons during locomotion. Brain Res 46:85—98

Orlovsky GN (1972c) Activity of rubrospinal neurons during locomotion. Brain Res 46:99—112

Orlovsky GN (1972d) Work of the Purkinje cells during locomotion. Biophysic 17:935—941

Orlovsky GN (1972e) Work of the neurones of the cerebellar nuclei during locomotion. Biophysics 17:1177—1185

Orlovsky GN, Feldman AG (1972a) On the role of afferent activity in generation of stepping movements (in Russian). Neirofiziologya 4:401—409

Orlovsky GN, Feldman AG (1972b) Classification of lumbosacral neurones according to their discharge patterns during evoked locomotion (in Russian). Neirofiziologya 4:410—417

Orlovsky GN, Pavlova GA (1972a) Vestibular responses of neurons of different descending pathways in cats with intact cerebellum and in decerebellated ones (in Russian). Neirofiziologya 4:303—310

Orlovsky GN, Pavlova GA (1972b) Vestibular responses in neurons of descending pathways during locomotion (in Russian). Neirofiziologya 4:311—317

Orlovsky GN, Pavlova GA (1972c) Response of Deiters' neurons to tilt during locomotion. Brain Res 42:212—214

Orlovsky GN, Shik ML (1976) Control of locomotion: a neuro-physiological analysis of the cat locomotor system. In: Internat Rev Physiol Neurophysiology II, vol 10, pp 281—317

Oscarsson O (1957) Functional organization of the ventral spino-cerebellar tract in the cat. II. Connections with muscle, joint, and skin nerve afferents and effects on adequate stimulation of various receptors. Acta Physiol Scand, 42, Suppl 146:1—107

Oscarsson O (1960) Functional organization of the ventral spino-cerebellar tract in the cat. III. Supraspinal control of VSCT units of I type. Acta Physiol Scand 49:171—183

Oscarsson O (1965) Functional organization of the spino- and cuneocerebellar tracts. Physiol Rev 45:495—522

Oscarsson O (1967) Functional significance of information channels from the spinal cord to the cerebellum. In: Purpura DP, Yahr MD (eds) Neurophysiological basis of normal and abnormal motor activities. 3rd Symposium of the Parkinson's Disease Information Center. Raven, New York, pp 93—117

Oscarsson O (1969a) The sagittal organization of the cerebellar anterior lobe as revealed by the projection patterns of the climbing fiber system. In: Llinás R (ed) Neurobiology of cerebellar evolution and development. American Medical Association, Chicago, pp 525—537

Oscarsson O (1969b) Termination and functional organization of the dorsal spino-olivocerebellar path. J Physiol (Lond) 200:129—149

Oscarsson O (1973) Functional organization of spinocerebellar paths. In: Iggo A (ed) Handbook of sensory physiology, vol II. Springer, Berlin Heidelberg New York, pp 339—380

Oscarsson O, Rosén I (1966) Response characteristics of reticulo-cerebellar neurones activated from spinal afferents. Exp Brain Res 1:320—328

Oscarsson O, Sjölund B (1977a) The ventral spino-olivocerebellar system in the cat. I. Identification of five paths and their termination in the cerebellar anterior lobe. Exp Brain Res 28:469—486

Oscarsson O, Sjölund B (1977b) The ventral spino-olivocerebellar system in the cat. II. Termination zones in the cerebellar posterior lobe. Exp Brain Res 28:487–503

Oscarsson O, Sjölund B (1977c) The ventral spino-olivocerebellar system in the cat. III. Functional characteristics of the five paths. Exp Brain Res 28:505–520

Otero JB (1976) Comparison between red nucleus and precentral neurons during learned movements in the monkey. Brain Res 101:37–46

Padel Y, Armand J, Smith AM (1972) Topography of rubrospinal units in the cat. Exp Brain Res 14:363–371

Padel Y, Steinberg R (1978) Red nucleus cell activity in awake cats during a placing reaction. J Physiol (Paris) 74:265–282

Palay S, Chan-Palay V (1974) Cerebellar cortex cytology and organization. Springer, Berlin Heidelberg New York

Palkovits M, Magyar P, Szentágothai J (1971a) Quantitative histological analysis of the cerebellar cortex in the cat. I. Number and arrangement in space of the Purkinje cells. Brain Res 32:1–13

Palkovits M, Magyar P, Szentágothai J (1971b) Quantitative histological analysis of the cerebellar cortex in the cat. II. Cell number and densities in the granular layer. Brain Res 32:15–30

Palkovits M, Magyar P, Szentágothai J (1971c) Quantitative histological analysis of the cerebellar cortex in the cat. III. Structural organization of the molecular layer. Brain Res 34:1–18

Palkovits M, Magyar P, Szentágothai J (1972) Quantitative histological analysis of the cerebellar cortex in the cat. IV. Mossy fiber-Purkinje cell numerical transfer. Brain Res 45:15–29

Palkovits M, Mezey E, Hamori J, Szentágothai J (1977) Quantitative histological analysis of the cerebellar nuclei in the cat. I. Numerical data on cells and on synapses. Exp Brain Res 28:189–209

Panchin YuV (1978) Vestibular responses of fast and slow Deiters' neurons (in Russian). Neirofiziologya 10:313–315

Pauls AR, Pauls RM, Soye I, Stein JF (1974) The activity of cerebellar neurones during manual tracking in awake rhesus monkeys. J Physiol (Lond) 242:64

Pavlova GA (1977) Activity of reticulo-spinal neurons during scratching (in Russian). Biofizika 22:740–742

Pearson KG, Duysens J (1976) Function of segmental reflexes in the control of stepping in cockraches and cats. In: Herman RM, Grillner S, Stein P, Stuart D (eds) Neural Control of Locomotion. Plenum, New York, pp 519–537

Perret C (1973) Analyse des mécanismes d'une activité de type locomoteur chez le chat. Thèse Doct Sci, Paris, CNRS AO 8342

Perret C (1976) Neural control of locomotion in the decorticate cat. In: Herman RM, Grillner S, Stein PSG, Stuart DG (eds) Neural control of locomotion. Plenum, New York, pp 587–615

Perret C (1983) Centrally generated pattern of motoneuron activity during locomotion in the cat. In: Roberts A, Roberts B (eds) Neural origin of rhythmic movement, pp 405–422

Perret C, Berthoz A (1973) Evidence of static and dynamic fusimotor actions on the spindle response to sinusoidal stretch during locomotor activity in the cat. Exp Brain Res 18:178–188

Perret C, Buser P (1972) Static and dynamic fusimotor activity during locomotor movements in the cat. Brain Res 40:165–169

Perret C, Cabelguen J-M (1976) Central and reflex participation in the timing of locomotor activations of a bifunctional muscle, the semitendinosus, in the cat. Brain Res 106:390–395

Perret C, Cabelguen J-M (1980) Main characteristics of the hindlimb locomotor cycle in the decorticate cat with special reference to bifunctional muscles. Brain Res 187:333–352

Perret C, Millanvoye M, Cabelguen J-M (1972) Messages spinaux ascendants, pendant une locomotion fictive chez le chat curarisé. J Physiol (Paris) 65:153A

Peterson BW (1979) Reticulospinal projections to spinal motor nuclei. Annu Rev Physiol 41:127–140

Peterson BW, Felpel LP (1971) Excitation and inhibition of reticulospinal neurons by vestibular, cortical and cutaneous stimulation. Brain Res 27:373–376

Peterson BW, Anderson ME, Filion M (1974) Responses of ponto-medullary reticular neurons to cortical, tectal and cutaneous stimuli. Exp Brain Res 21:19–44

Peterson BW, Maunz RA, Pitts NG, Mackel RG (1975) Patterns of projection and branching of reticulospinal neurons. Exp Brain Res 23:333–351

Peterson BW, Maunz RA, Fukushima K (1978a) Properties of a new vestibulospinal projection, the caudal vestibulospinal tract. Exp Brain Res 32:287–292

Peterson BW, Pitts NG, Fuskushima K, Mackel R (1978b) Reticulospinal excitation and inhibition of neck motoneurons. Exp Brain Res 32:471–489

Peterson BW, Pitts NG, Fuskhima K (1979) Reticulospinal connections with limb and axial motoneurones. Exp Brain Res 36:1–20

Petras JM (1967) Cortical, tectal and tegmental fiber connections in the spinal cord of the cat. Brain Res 6:275–324

Philippson M (1905) L'autonomie et la centralisation dans le système nerveux des animaux. Trav Lab Physiol Inst. Solvay, Bruxelles 7:1–208

Pompeiano O (1959) Organizzazione somatotopica delle risposte flessorie alla stimulazione elettrica del nucleo interposito nel gatto decerebrato. Arch Sci Biol 43:173–176

Pompeiano O (1960) Organizzazione somatotopica delle risposte posturali alla stimulazione elettrica del nucleo di Deiters nel gatto decerebrato. Arch Sci Biol 44:497–511

Pompeiano O (1967) Functional organization of the cerebellar projections to the spinal cord. In: Fox CA, Snider RS (eds) Progress in brain research, vol 25, The cerebellum. Elsevier, Amsterdam, pp 282–321

Pompeiano O (1973) Reticular formation: In: Iggo A (ed) The somatosensory system. Springer, Berlin Heidelberg New York, pp 381–488

Pompeiano O (1975a) Macular input to neurons of the spinoreticulocerebellar pathway. Brain Res 95:351–368

Pompeiano O (1975b) Vestibulo-spinal relationships. In: Naunton RF (ed) The vestibular system. Academic, New York, pp 147–184

Pompeiano O (1977) Macular influences on somatosensory transmission through the spinoreticulocerebellar pathway. J Physiol (Paris) 73:387–400

Pompeiano O, Barnes CD (1971) Response of brain stem reticular neurons to muscle vibration in the decerebrate cat. J Neurophysiol 34:709–724

Pompeiano O, Brodal A (1957a) The origin of vestibulospinal fibres in the cat. An experimental anatomical study with comments on the descending medial longitudinal fasciculus. Arch Ital Biol 95:166–195

Pompeiano O, Brodal A (1957b) Experimental demonstration of a somatotopical origin of rubrospinal fibres in the cat. J Comp Neurol 108:225–252

Pompeiano O, Cotti E (1959) Analisi microelettrodica della proiezioni cerebello-deitersinae. Arch Sci Biol 43:57–101

Prochazka A, Westerman RA, Ziccone SP (1976) Discharges of single hindlimb afferents in the freely moving cat. J Neurophysiol 39:1090–1104

Prochazka A, Westerman RA, Ziccone SP (1977) Ia afferent activity during a variety of voluntary movements in the cat. J Physiol (Lond) 268:423–448

Pyatigorsky BYa (1968) The dynamics of the synaptic transmission of the cutaneous and proprioceptive impulsation on some spinal neurons. In: Kostyuk PG (ed) Synaptic processes (in Russian). Kiev, pp 272–291

Ramón y Cajal S (1911) Histologie du système nerveux de l'homme et des vertébrés. T II, A Maloine, Paris

Rasmussen S, Chan AK, Goslow GE (1978) The cat step cycle: Electromyographic patterns for hindlimb muscles during posture and unrestrained locomotion. J Morphol 155:253–270

Reighard J, Jennings HS (1935) Anatomy of the cat. Henry Holt, New York

Rexed B (1954) A cytoarchitectonic atlas of the spinal cord in the cat. J Comp Neurol 100:297–379

Rhines R, Magoun HW (1946) Brain stem facilitation of cortical motor response. J Neurophysiol 9:216–229

Robertson LT, Grimm RJ (1975) Responses of primate dentate neurons to different trajectories of the limb. Exp Brain Res 23:447–462

Romanes GJ (1964) The motor pools of the spinal cord. In: Eccles JC, Schade JP (eds) Progress in brain research, vol 2, Organization of the spinal cord. Elsevier, Amsterdam, pp 93–119

Rosén I, Scheid P (1972) Cerebellar surface cooling influencing evoked activity in cortex and in interpositus nucleus. Brain Res 45:580–584

Rosén I, Scheid P (1973a) Patterns of afferent input to the lateral reticular nucleus of the cat. Exp Brain Res 18:242–255

Rosén I, Scheid P (1973b) Responses to nerve stimulation in the bilateral ventral flexor reflex tract of the cat. Exp Brain Res, 18:256–267

Rosén I, Scheid P (1973c) Responses in the spino-reticulocerebellar pathway to stimulation of cutaneous mechanoreceptors. Exp Brain Res 18:268–277

Rossi GF, Brodal A (1956) Corticofugal fibres to the brainstem reticular formation. An experimental study in the cat. J Anat (Lond) 90:42–62

Rossi GF, Zanchetti A (1957) The brain stem reticular formation. Anatomy and physiology. Arch Ital Biol 95:203–435

Rushmer DS, Roberts WJ, Augter GK (1976) Climbing fiber responses of cerebellar Purkinje cells to passive movement of the cat fore paw. Brain Res 106:1–20

Russell DF, Zajac FE (1979) Effects of stimulating Deiters' nucleus and medial longitudinal fasciculus on the timing of the fictive locomotor rhythm induced in cats by DOPA. Brain Res 177:588–592

Sasaki K, Namikawa A, Harshiramoto S (1960) The effect of midbrain stimulation upon alpha motoneurones in lumber spinal cord of the cat. Jpn, J Physiol 10: 303–316

Sasaki K, Tanaka T, Mori K (1962) Effects of stimulation of pontine and bulbar reticular formation upon spinal motoneurones of the cat. Jpn, J Physiol 12:43–62

Schor RH (1974) Responses of vestibular neurones to sinusoidal roll tilt. Exp Brain Res 20:347–362

Schultz W, Montgomery EB, Marini R (1976) Sterotyped flexion of forelimb and hindlimb to microstimulation of dentate nucleus in cebus monkeys. Brain Res 107:151–155

Schultz W, Montgomery EB, Marini R (1979) Proximal limb movements in response to microstimulation of primate dentate and interpositus nuclei mediated by brain-stem structures. Brain 102:127–146

Sedgwick EM, Williams TDW (1967) Responses of single units in the inferior olive to stimulation of the limb nerves, peripheral skin receptors, cerebellum, caudate nucleus and motor cortex. J Physiol (Lond) 189:261–279

Selverston AI (1980) Are central pattern generators understandable? Behavioral Brain Sci 3:535–571

Severin FV (1970) On the role of $\gamma$-motor system for extensor $\alpha$-motoneuron activation during controlled locomotion. Biophysics 15:1138–1145

Severin FV, Orlovsky GN, Shik ML (1967a) Work of the muscle receptors during controlled locomotion. Biophysics 12:575–586

Severin FV, Shik ML, Orlovsky GN (1967b) Work of the muscles and single motoneurones during controlled locomotion. Biophysics 12:762–772

Shapovalov AI (1975a) Neuronal organization and synaptic mechanisms of supraspinal motor control in vertebrates. Rev Physiol Biochem Pharmacol 72:1–54

Shapovalov AI (1975b) Neurones and synapses of supraspinal motor systems (in Russian). Nauka, Leningrad

Shapovalov AI, Shapovalova KB (1966) Activity of alpha-motoneurones during rhythmical stimulation of red nucleus and effects of strychnine on rubrospinal influences (in Russian). Dokl Acad Nauk SSSR, 168:1430–1433

Shapovalov AI, Grantyn AA, Kurchavyi GG (1967) Short-latency reticulospinal projections to alpha-motoneurones (in Russian). Bull Exp Biol Med 64:3–9

Shapovalov AI, Karamjan OA, Tamarova ZA, Kurchavyi GG (1972) Cerebello-rubrospinal effects on hindlimb motoneurones in the monkey. Brain Res 47: 49–59

Sherrington CS (1898) Decerebrate rigidity and reflex coordination of movements. J Physiol (Lond) 22:319–332

Sherrington CS (1906a) The integrative action of the nervous system. Yale University Press, New Haven

Sherrington CS (1906b) Observations on the scratch-reflex in the spinal dog. J Physiol (Lond) 34:1–50

Sherrington CS (1910a) Flexion-reflex of the limb, crossed extension reflex, and stepping and standing. J Physiol (Lond) 40:28–121

Sherrington CS (1910b) Notes on the scratch reflex of the cat. Q J Exp Physiol 3:213–220

Sherrington CS (1917) Reflexes elicitable in the cat from pinna, vibrissae and jaws. J Physiol (Lond) 51:404–431

Shik ML, Orlovsky GN (1976) Neurophysiology of locomotor automation. Physiol Rev 56:465–501

Shik ML, Orlovsky GN, Severin FV (1966a) Organization of locomotor synergism. Biophysics 11:1011–1019

Shik ML, Severin FV, Orlovsky GN (1966b) Control of walking and running by means of electrical stimulation of the midbrain. Biophysics 11:756–765

Shik ML, Severin FV, Orlovsky GN (1967) Structures of the brain stem responsible for evoked locomotion (in Russian). Fiziol Zh SSSR, 12:660–668

Shimamuza M, Kagure J, Wada S-J (1982) Reticular neuron activities associated with locomotion in thalamic cats. Brain Res 231:51–62

Shurrager PS (1955) Walking in spinal kittens and puppies. In: Windle WF (ed) Regeneration in the central nervous system. Ch C Thomas, Springfield, pp 208–218

Sjölund B (1978) The ventral spino-olivocerebellar system in the cat. V. Supraspinal control of spinal transmission. Exp Brain Res 33:509–522

Sjöström A, Zangger P (1976) Muscle spindle control during locomotor movements generated by the deafferented spinal cord. Acta Physiol Scand 97:281–291

Smirnov KA, Potechina IL (1974) Localization and properties of reticulo-spinal neurons with axons descending in the dorsolateral parts of the spinal cord lateral funiculi (in Russian). Neirofiziologya, 6:266–272

Smith AM, Bourbonnais D (1981) Neuronal activity in cerebellar cortex related to control of prehensile force. J Neurophysiol 45:286–303

Smolyaninov VV (1966) Some special features of organization of the cerebellar cortex. In: Gelfand IM, Gurfinkel VS, Fomin SV, Tsetlin ML (eds) Models of the structural-functional organization of certain biological systems (in Russian). Nauka, Moscow, pp 203–262
(English translation by MIT Press, Massachusetts 1971)

Snider RS (1940) Morphology of the cerebellar nuclei in rabbit and cat. J Comp Neurol 72:399–415

Snider RS (1952) Interrelations of cerebellum and brain stem. A Res Nerv Ment Dis Proc 30:267–281

Snider RS, Stowell A (1944) Receiving areas of the tactile, auditory and visual systems in the cerebellum. J Neurophysiol 7:331–357

Soechting JE, Burton JE, Onoda N (1978) Relationships between sensory input, motor output and unit activity in interpositus and red nuclei during intentional movement. Brain Res 152:65−79

Sprague JM, Chambers WW (1953) Regulation of posture in intact and decerebrate cat. I. Cerebellum, reticular formation, vestibular nuclei. J Neurophysiol 16: 451−463

Sprague JM, Chambers WW (1954) Control of posture by reticular formation and cerebellum in the intact, anesthetized and unanesthetized and in the decerebrated cat. Am J Physiol 176:52−64

Sprague JM, Schreiner LH, Lindsley DB, Magoun HW (1948) Reticulo-spinal influences on stretch reflexes. J Neurophysiol 11:501−507

Steeves JD, Jordan LM (1980) Localization of a descending pathway in the spinal cord which is necessary for controlled treadmill locomotion. Neurosci Lett 20:283−288

Stein JF (1978) Long loop motor control in monkeys. The effects of transient cooling of parietal cortex and of cerebellar nuclei during tracking tasks. In: Desmedt JE (ed) Progress in clinical neurophysiology, vol 4, cerebral motor control in man. S Karger, Basel, pp 107−122

Stein PG (1978) Motor systems, with specific reference to the control of locomotion. Annu Rev Neurosci 1:61−81

Strick PL (1983) The influence of motor preparation on the response of cerebellar neurons to limb displacements. J Neurosci 3:2007−2020

Sumi T (1969) Some properties of cortically-evoked swallowing and chewing in rabbits. Brain Res 15:107−120

Sumi T (1970) Activity in single hypoglossal fibers during cortically induced swallowing and chewing in rabbits. Pflügers Archiv Gesamte Physiol 314:329−346

Szentagothai J (1964) Anatomical aspects of junctional transformation. In: Gerard RW, Duyff JW (eds) Information processing in the nervous system, vol 3. Excerpta Medica Foundation, Amsterdam, pp 119−136

Szentagothai J, Albert A (1955) The synaptology of Clarke's column. Acta Morphol Hung 5:43−51

Szentagothai J, Rajkovits K (1959) Über den Ursprung der Kletterfasern des Kleinhirns. Z Anat Entwicklungsgesch 121:130−141

Terzuolo CA (1959) Cerebellar inhibitory and excitatory actions upon spinal extensor motoneurons. Arch Ital Biol 97:316−339

Thach WT (1967) Somatosensory receptive fields of single units in cat cerebellar cortex. J Neurophysiol 30:675−696

Thach WT (1968) Discharge of Purkinje and cerebellar nuclear neurons during rapidly alternating arm movements in the monkey. J Neurophysiol 31:785−797

Thach WT (1970a) Discharge of cerebellar neurons related to two maintained postures and two prompt movements. I. Nuclear cell output. J Neurophysiol 33:527−536

Thach WT (1970b) Discharge of cerebellar neurons related to two maintained postures and two prompt movements. II. Purkinje cell output and input. J Neurophysiol 33:537−547

Thach WT (1972) Cerebellar output: properties, synthesis and uses. Brain Res 40:89−97

Thach WT (1975) Timing of activity in cerebellar dentate nucleus and cerebral motor cortex during prompt volitional movement. Brain Res 88:233−241

Thach WT (1978a) Correlation of neural discharge with pattern and force of muscular activity, joint position, and direction of intended next movement in motor cortex and cerebellum. J Neurophysiol 41:654−676

Thach WT (1978b) Single unit studies of long loops involving the motor cortex and cerebellum during limb movements in monkeys. In: Desmedt JE (ed) Progress clinical neurophysiology, vol 4, cerebral motor control in man. S Karger, Basel, pp 94−106

Thulin C-A (1953) Motor effects from stimulation of the vestibular nuclei and the reticular formation. Acta Physiol Scand, 28, Suppl 103:1—61

Thulin C-A (1963) Effects of electrical stimulation of the red nucleus on the alpha motor system. Exp Neurol 7:464—484

Tolbert DL, Bantli H, Hames EG, Ebner TJ, McMullen TA, Bloedel JR (1980) A demonstration of the dentato-reticulospinal projection in the cat. Neuroscience 5:1479—1488

Torvik A, Brodal A (1957) The origin of reticulospinal fibers in the cat. An experimental study. Anat Rec 128:113—137

Toyama K, Tsukahara N, Udo M (1968) Nature of the cerebellar influences upon the red nucleus. Exp Brain Res 4:292—309

Toyama K, Tsukahara N, Kosaka K, Matsunami K (1970) Synaptic excitation of red nucleus neurones by fibres from interpositius nucleus. Exp Brain Res 11: 187—198

Trouche E, Beaubaton D (1980) Initiation of a goal-directed movement in the monkey. Exp Brain Res 40:311—321

Tsukahara N, Toyama K, Kosaka K (1964) Intracellularly recorded responses of red nucleus neurones during antidromic and orthodromic activation. Experientia (Basel), 20:632—633

Tsukahara N, Toyama K, Kosaka K, Udo M (1965) "Disfacilitation" of red nucleus neurones. Experientia (Basel), 21:544—545

Tsukahara N, Toyama K, Kosaka K (1967) Electrical activity of the red nucleus neurones investigated with intracellular microelectrodes. Exp Brain Res 4: 18—33

Udo M (1975) Cerebellar control of walking investigated in cat Deiters' neurones. J Physiol Soc Jpn, 37:380—381

Udo M, Mano N (1970) Discrimination of different spinal monosynaptic pathways converging onto reticular neurones. J Neurophysiol 33:227—238

Udo M, Oda Y, Tanaka K, Horikawa J (1976) Cerebellar control of locomotion: discharge from Deiters' neurones, EMG and limb movements during local cooling of the cerebellar cortex. In: Homma S (ed) Progress brain research, vol 44, Understanding the stretch reflex. Elsevier, Amsterdam, pp 445—459

Udo M, Matsukawa K, Kamei H (1979a) Effects of partial cooling of cerebellar cortex at lobules V and IV of the intermediate part in the decerebrate walking cats under monitoring vertical floor reaction forces. Brain Res 160:559—564

Udo M, Matsukawa K, Kamei H (1979b) Hyperflexion and changes in interlimb coordination of locomotion induced by cooling of the cerebellar intermediate cortex in normal cats. Brain Res 166:405—408

Udo M, Matsukawa K, Kamei H, Oda Y (1980) Cerebellar control of locomotion: effects of cooling cerebellar intermediate cortex in high decerebrate and awake walking cats. J Neurophysiol 44:119—134

Udo M, Matsukawa K, Kamei H, Minoda K, Oda Y (1981) Simple and complex spike activities of Purkinje cells during locomotion in the cerebellar vermal zones of decerebrate cats. Exp Brain Res 41:292—300

Uno M, Kozlovskaya I, Brooks VB (1973) Effects of cooling interposed nuclei on tracking-task performance in monkeys. J Neurophysiol 36:996—1003

Vasilenko DA, Zadorozhny AG, Kostyuk PG, Pyatigorsky BYa (1969) Corticofugal postsynaptic influences on the neurons of Clarke's column (in Russian). Neirofiziologya, 1:15—24

Viala D, Buser P (1969) The effect of DOPA and 5-HTP on rhythmic efferent discharges in hindlimb nerves in the rabbit. Brain Res 12:437—443

Viala D, Vidal C (1978) Evidence for distinct spinal locomotion generators supplying respectively fore- and hindlimbs in the rabbit. Brain Res 155:182—186

Viala G, Coston A, Buser P (1970) Participation de cellules du cortex cérébelleux aux rythmes "locomoteurs" chez le lapin curarisé en absence d'informations somatiques liees au movement. C R Acad Sci Paris, Sér D, 271:688—691

Viala D, Valin A, Buser P (1974) Relationship between the "late reflex discharge" and locomotor movements in acute spinal cats and rabbits treated with DOPA. Arch Ital Biol 112:299–306

Vidal C, Viala D, Buser P (1979) Central locomotor programming in the rabbit. Brain Res 168:57–73

Voogd J (1964) The cerebellum of the cat. Structure and fibre connexions. Van Gorcum, Assen

Walberg F, (1958) On the termination of the rubrobulbar fibers. Experimental observations in the cat. J Comp Neurol 110:65–73

Walberg F, Jansen J (1961) Cerebellar corticovestibular fibers in the cat. Exp Neurol 3:32–52

Walberg F, Jansen J (1964) Cerebellar corticonuclear projection studied experimentally with silver impregnation methods. Z Hirnforsch 6:338–354

Walberg F, Pompeiano O (1960) Fastigio-fugal fibers to the lateral reticular nucleus. An experimental study in the cat. Exp Neurol 2:40–53

Walberg F, Westrum LE, Hauglie-Hanssen E (1962) Fastigioreticular fibers in the cat. An experimental study with silver method. J Comp Neurol 119:187–199

Wetzel MC, Stuart DG (1976) Ensemble characteristics of cat locomotion and its neural control. Prog Neurobiol 7:1–98

Wetzel MC, Atwater AE, Wait JV, Stuart DG (1975) Neural implications of different profiles between treadmill and overground locomotion timings in cats. J Neurophysiol 38:492–501

Wilson VJ (1972) Physiological pathways through the vestibular nuclei. Int Rev Neurobiol 15:27–81

Wilson VJ, Peterson BW (1978) Peripheral and central substrates of vestibulospinal reflexes. Physiol Rev 58:80–105

Wilson VJ, Yoshida M (1969) Comparison of effects of stimulation of Deiters' nucleus and medial longitudinal fasciculus on neck, forelimb, and hindlimb motoneurons. J Neurophysiol 32:743–758

Wilson VJ, Kato M, Thomas RC, Peterson BW (1966) Excitation of lateral vestibular neurons by peripheral afferent fibers. J Neurophysiol 29:508–529

Wilson VJ, Kato M, Peterson BW, Wylie RM (1967) A single-uni analysis of the organization of Deiters' nucleus. J Neurophysiol 30:603–619

Wilson VJ, Uchino Y, Susswein A, Fukushima K (1977) Properties of direct fastigiospinal fibers in the cat. Brain Res 126:543–546

Wilson VJ, Uchino Y, Maunz RA, Susswein A, Fukushima K (1978) Properties and connections of cat fastigiospinal neurons. Exp Brain Res 32:1–18

Wolstencroft JN (1964) Reticulospinal neurones. J Physiol (Lond) 174:91–108

Wylie RM, Felpel LP (1971) The influence of the cerebellum and peripheral somatic nerves on the activity of Deiters' cells in the cat. Exp Brain Res 12:528–546

Wyman RJ (1976) Neurophysiology of the motor output pattern generator for breathing. Fed Proc 35:2013–2023

Yamaguchi T (1982) Activities of forelimb motoneurones during fictive stepping in decerebrate cats. J Physiol Soc Jpn 44:387

Yamamoto K, Odagiri M (1981) Discharge pattern differences between cat interpositus and dentate neurons during isometric lever pressing. Exp Brain Res 43:104–106

Yu J (1972) The pathway mediating ipsilateral limb hyperflexion after cerebellar paravermal cortical ablation or cooling in cats. Exp Neurol 36:549–562

Yu J, Eidelberg E (1981) Effects of vestibulospinal lesions upon locomotor function in cats. Brain Res 220:179–183

Yu J, Tarnecki R, Chambers WW, Liu CN, Konorski J (1973) Mechanisms mediating ipsilteral limb hyperflexion after cerebellar paravermal cortical ablation or cooling. Exp Neurol 38:144–156

Zajac FE, Young JL (1980) Discharge properties of hindlimb motoneurons in decerebrate cats during locomotion induced by mesencephalic stimulation. J Neurophysiol 43:1221–1235

Zangger P (1978) Fictive locomotion in curarized high spinal cats elicited with 4-aminophyridine and DOPA. Experientia (Basel) 34:904

Zangger P (1981) The effect of 4-aminopyridine on the spinal locomotor rhythm

# Subject Index